WHAT'S
MY CAT
THINKING?

WHAT'S MY CAT THINKING?

UNDERSTAND YOUR CAT TO GIVE THEM A HAPPY LIFE

DR. JO LEWIS

DK | Penguin Random House

Project Editor Zia Mattocks
Project Art Editor Karen Constanti
Senior Editor Rona Skene
US Editor Megan Douglass
US Consultant Beth Adelman, MS
Editorial Assistant Kiron Gill
Project Designer Louise Brigenshaw
Managing Editor Dawn Henderson
Managing Art Editor Marianne Markham
Production Editor David Almond
Production Controller Luca Bazzoli
Jacket Designer Amy Cox
Jacket Coordinator Lucy Philpott
Art Director Maxine Pedliham
Publishing Director Katie Cowan

Illustrations Mark Scheibmayr

First American Edition, 2021
Published in the United States by DK Publishing
1450 Broadway, Suite 801, New York, NY 10018

21 22 23 24 25 10 9 8 7 6 5 4 3 2
006-323486-Sept/2021

Published in Great Britain by Dorling Kindersley Limited

A catalog record for this book is available from the Library of Congress.
ISBN: 978-0-7440-3985-6

Printed and bound in China.

DISCLAIMER see page 191

For the curious

www.dk.com

Contents

MY CAT AND ME

MY CAT DRIVES ME CRAZY!

WHAT'S UP WITH MY CAT?

ADVANCED CAT WATCHING

SURVIVAL GUIDES

Foreword

I'm an unashamed crazy cat lady. There, I've said it! Born into a family of cat lovers and with decades of experience as a cat vet, I've spent my whole life loving and learning about cats. The raw deal they often get when visiting the vet inspired me to found The Cat Vet and the UK's first dedicated feline home-visiting vet clinic. Every day I see the intimate connection between a cat's mental and physical health, which is perhaps why the feline brain remains so fascinating to me.

Cats are beautifully complex creatures—our affectionate companions, but always in touch with their wild side. That's part of their appeal, yet it's also why they struggle at times to fit into our modern homes and lifestyles. They don't convey their thoughts and emotions like we do, and it's their instinct to conceal any signs of ill health, so their quiet discontent can easily go unnoticed.

The Internet proves we're intrigued by cats—yet every year millions are abandoned for not meeting human expectations. Let's be honest, we all talk to our cats—but we need to listen more to what they're telling us. I passionately believe this is the key to understanding them and improving their health and happiness.

Seeing and treating cats in their familiar habitat has given me a unique insight into their behavior, and I leapt at the chance to share some of the science behind what I've learned. This book is the result, a fascinating look at common cat behaviors that people find amusing, baffling, frustrating, or worrying (or perhaps should!).

I hope you'll enjoy putting your newfound "cat watching" skills into practice, and will start to see cats in a different light and with a deeper understanding. If we all stop to ponder "What's my cat thinking?" felines of the future will stand a better chance of being treated with the respect and compassion that all cats deserve.

INTRODUCTION

Think like a cat

Interpreting what your cat might be thinking means understanding their perspective, motivations, and subtle communication style—all determined by a complex mix of primal instincts, genetics, and learned behavior.

The wildcat within

Our modern feline companions are extremely close genetically to the wildcats they descended from. They might strut around our living rooms in their designer fur coats, but they're still wild at heart. So trying to make heads or tails of their thoughts and behavior involves viewing their world through a wildcat lens.

As you read through this book, you will discover the underlying instincts and motivations behind some of the behaviors I am asked about most on my rounds as a cat vet. You will see lots of mentions of our pet cats inheriting wildcat traits, such as being solitary, territorial hunters. There is a misconception that this means they are fierce and robust, but they're not. All cats are vulnerable—they are prey for bigger carnivores, super sensitive (see pages 14–15), and often far less chilled-out than we think they are (see pages 122–123). Our modern pet cats still have wildcat urges and needs. Realizing this helps us understand what they might be thinking. Only then can we hope to offer them a long, happy, and healthy life with us.

Territorial

A wildcat's patch of earth and all it contains is what ensures their survival. It really is their whole world, so they're prepared to fight tooth and claw for it. Pet cats are more tolerant and sociable than their ancestors, but whether they're willing to share their territory is down to the individuals involved; the space being offered; and the number, type, and location of resources. The smaller the space or the more cats you have, the lower the odds of harmony and the more stressed a cat can become.

Solitary survivor

Our cat's ancestral wildcat was a free spirit, instinctively geared to be master of their own destiny. Similarly, our pet cats really value control, choice, and a predictable routine, and they struggle with sharing or compromise. They're very set in their ways, and are happiest when they can choose where they live and do what they like, when they like.

Hunter and hunted

On a sofa or in the savanna, a cat's eyes, ears, and whiskers are wired 24-7, scanning for prey and predators. Anything big, noisy, fast-moving, or unpredictable triggers most cats' instinct to escape and either lie low or perch high while they gather information about the threat. Self-reliant by necessity, they are always looking—and sniffing—over their shoulder for signs of change.

Active mind, active body

With a rodent quota to fill to keep them sustained, wildcats have to be curious, attentive, fit, and persistent. Similarly, pet cats are expert explorers, foragers, and problem-solvers. Most are natural acrobats and energetic climbers—though their curved claws mean they have yet to master graceful descents!

Making your cat feel at home

We might offer cats a sheltered existence in our homes, but we have to provide enough 3D space and vital resources in that territory to keep them active, happy, and healthy (see pages 46–47). Ensuring they don't feel bored or threatened also helps avoid stress-induced diseases (see 164–165). It's often only when stresses build up and tip them over the edge that a cat's inner turmoil gets noticed (see pages 30–31). We might see it as a change in litter box or eating habits, grumpiness, or a change in activity levels.

African wildcat

Genetically, pet cats are not that far removed from their wildcat ancestors, whose natural drives and instincts underpin many feline behavioral traits.

How cats communicate

Cats make their needs, desires, and feelings known to us and fellow felines all the time in a variety of ways, and our job is to find ways to decipher what they mean. All you need are keen powers of observation, a willingness to learn to interpret their body language and sounds, and an understanding of the fascinating world of scent.

Body language

Analyzing how a cat rests; what their tail does; and how their eyes, ears, and whiskers look helps you avoid making assumptions about their mood. Each part of their body reveals a piece of the picture, and recognizing those individual signals and putting them together gives you a snapshot of what your cat is telling you in that moment.

Begin by studying different parts of your cat's body and monitor how they change in various situations. Always look at the whole cat and the context they're in. If you're perplexed by an odd mix of visual signals, your cat might be confused, unsure, or feeling multiple emotions at once.

Don't make assumptions

This typical cat greeting could easily be interpreted as a request for food, if they happen to display this body language when you are in the kitchen, but it could just be a plea for attention or some other unmet need.

Body

Is your cat's posture relaxed or tense? Are they standing tall, arching their back, or side-on to appear bigger; or are they crouching, slinking with their tummy to the floor, or hunched with their head pulled in to their body? Are paws grounded or relaxed? Are vulnerable areas (such as the tummy) hidden or on display? Is fur erect or skin twitching or rippling with tension? Remember not to leap to conclusions—a cat who is motionless while being stroked could be content and relaxed, or frozen with fear and wanting to hide (see pages 122–123).

Ears

Upright, forward-facing ears suggest a cat is relaxed and alert, or wanting to intimidate another cat. Flattened-to-the-side "airplane" ears signal fear and a keenness to back off. Rotated-back "Batman" ears flag agitation (see pages 102–103). Ears hovering between "airplane" and "Batman," or held so far back they seem to be missing suggest mixed emotions. An oscillating ear indicates they may be processing two sources of sound—left-ear movements imply it's a negative sound, regardless of direction. Ear twitches or flicks may signal agitation, anxiety, or an itch.

Eyes

Are your cat's eyes "hard," round, and focused, or "soft," almond-shaped, and relaxed? Are they making eye contact (confident or challenging) or averting their gaze (avoiding confrontation)? Fearful cats may look more to the left and relaxed cats more to the right. Adrenaline dilates the pupils and increases blink rate, whereas the pupils constrict when a cat is relaxed (or in bright light), and slow blinking shows contentment. Fully closed eyes suggests they're asleep, but may also be a response to sensory hyperstimulation, anxiety, or pain.

Whiskers

These super-sensitive feelers on the cheeks, muzzle, above the eyes, and on the backs of the front legs detect airflow and prey movement when navigating and hunting. They also alter their position with mood. Calm, sideways, slightly fanned whiskers become compact and flat against the cheeks with fear or frustration. Forward-swept whiskers can signal curiosity when they're curved and fanned, or pain if they're straight.

Tail

The tail is used for balance and also reveals a cat's mood (see page 27).

Continued 

Scent

A cat's superpower is their amazing sense of smell. Their nasal cavity has 40 times more smell receptors than ours, and the area of their brain concerned with olfaction (smelling) is much larger, too. A cat's brain recognizes thousands of scents and they're better than dogs at discriminating between smells—certainly not something to sniff at!

A cat's signature scent mix is their identity card, announcing their sex, age, family connections, reproductive status, health, mood, and more. Bodily excretions reveal what they've eaten, when they were last in the area, and which way they were going.

Scents are a lifeline for wildcats alone in vast territories, helping them

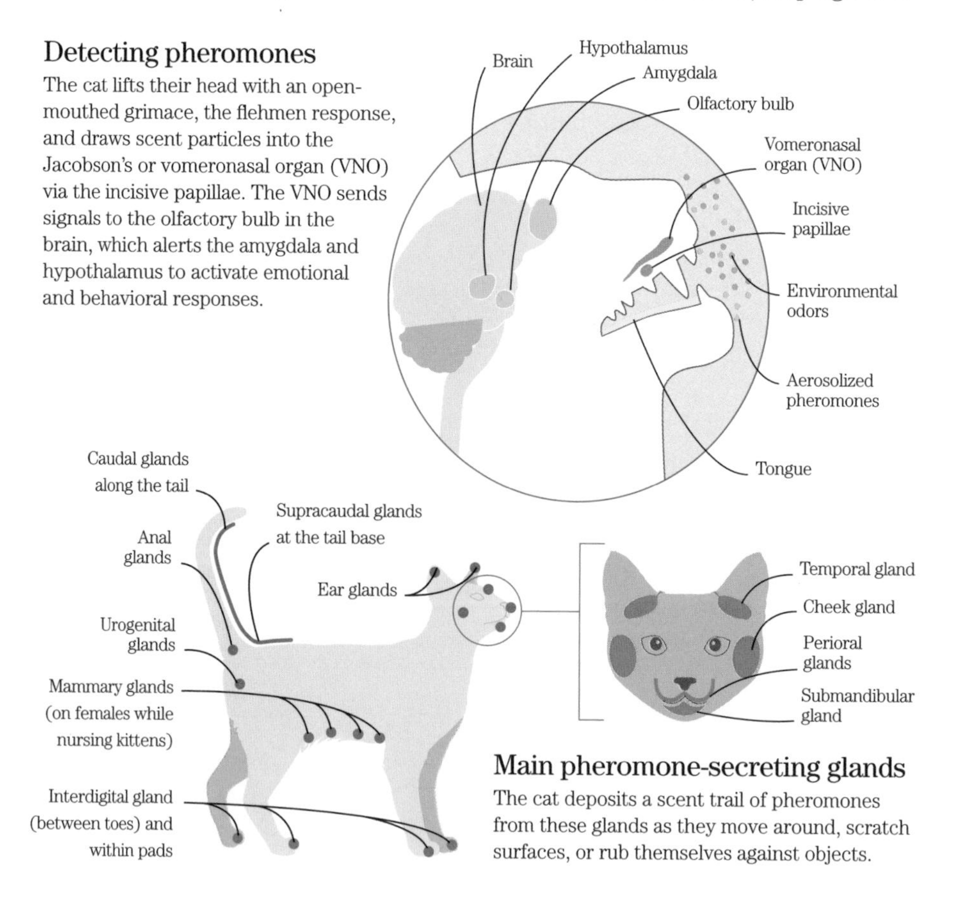

Detecting pheromones

The cat lifts their head with an open-mouthed grimace, the flehmen response, and draws scent particles into the Jacobson's or vomeronasal organ (VNO) via the incisive papillae. The VNO sends signals to the olfactory bulb in the brain, which alerts the amygdala and hypothalamus to activate emotional and behavioral responses.

Main pheromone-secreting glands

The cat deposits a scent trail of pheromones from these glands as they move around, scratch surfaces, or rub themselves against objects.

monitor their environment—from new mouse trails to a predator's poop. Lingering feline scents allow cats to chemically survey their space. Leaving a trail of scents sends other cats messages, avoids confrontations with rivals, and locates eligible mates.

Pheromones

Cats have an even more specialized sensory pathway for processing social information using pheromones—a species-specific scent. Released from scent glands located from head to tail (see opposite), pheromones are like an encrypted cat code that can signal moods and intentions, and trigger behavioral responses in other cats.

Relaxed cats rub friendly facial pheromones on surfaces, other cats, or trusted humans. Rubbing, rolling, and urine spraying can be sexual displays in unneutered cats, while spraying in neutered cats communicates anxiety or illness (see pages 120–121). Scratching their claws leaves both pheromones and a visual scratch mark to affirm their presence (see pages 134–135). A fearful cat's paws and anal glands expel alarm pheromones, while mother cats release appeasing pheromones to bond with their kittens.

How to make a face rub

Harvesting signature scents and friendly pheromones from cats helps make new or intimidating situations feel more familiar and reassuring. Scent rubs can be used to introduce new cats and kittens to your home (see pages 20–21) and new cats to existing pets (see pages 112–113).

- Use a clean piece of fabric, such as a thin cotton sock or glove, or a washcloth. First, wash it in enzyme detergent (ideally fragrance-free), double rinse, and let it dry.

- During a petting session, gently massage the fabric around the cat's facial scent glands (or nursing mother's mammary glands). Build up scent particles by repeating this process several times over several days—store the fabric in a resealable plastic bag in between sessions to marinate. Check for happy body language (see pages 82–83) during the session to ensure you're only collecting positive hormonal vibes and not alarm pheromones. Don't force things if your cat seems anxious or is not in the mood—a blanket they regularly sleep on is the next best thing to a scent rub.

- Just before the introduction, wipe the scent-loaded fabric onto the target item (such as the cat carrier, a new cat bed, new sofas, or door frames in a new house). Choose a height/location that mimics where a cat would naturally roll, rub, or rest.

Continued 

Sound

Cats have a vast vocal repertoire of different sounds that they use in various contexts to express themselves. This wide range of vocalizations is vital for managing hostile interactions with rival cats or predators, for reproductive behavior, and for mother cats to communicate with their kittens.

In the past, ancestral wildcats who were better at reading human cues and making their own needs understood would, no doubt, have been rewarded with tidbits. Today, a pet cat's everyday survival depends on the goodwill of us two-legged giants, so it pays to talk themselves up so they don't get overlooked.

What's your cat really saying?
Cats use their voice in many different ways: to announce their presence with friendly greetings, and to make requests, raise concerns, issue rebukes, or sound the alarm. Thinking about the following key aspects can help you interpret what your cat might be trying to tell you.

Context
Look at your cat's body language and see what it's telling you (see pages 12–13). What was happening in your cat's environment just before and just after they made the sound? Does that reveal anything about the outcome they may have been hoping for?

Feline expletives
Vocalizing with a fixed open mouth means "get lost!"—or risk unleashing grumpy cat (see pages 102–103).

Pitch

- **High pitch**—Attention-grabbing sounds; shrieks during conflict; requests such as meows; and solicitation purrs
- **Low pitch**—Loud, intimidating warnings; growls; and intimate, close-range purrs

Mouth

- **Closed**—Purr or chirrup
- **Fixed open**—More threatening vocalizations, such as a growl, snarl, hiss, or shriek
- **Open, but gradually closing**—Meow, caterwaul, yowl, or howl
- **Opening and closing**—Chatter

Feline vocalization types

Greetings and requests

Purr—A polite request for more affection when relaxed, or for comfort when scared, unwell, in pain, giving birth, or approaching death. Purring potentially has self-soothing and tissue-healing roles, too. Some cats add an extra high-pitched cry that presses our nurturing instinct button, not unlike the way a human baby's cry does.

Meow—A greeting or a request for human assistance, food, affection, or something else they desire. The delivery ranges from upbeat greetings, gentle reminders, borderline harassments, or final demands—such as "Feed-me-nowww!" Kittens produce a high-pitched mew. Exclamatory meows tend to announce a successful hunt (see pages 100–101). Silent meows are too high-pitched for our inferior ears to detect.

Yowl (aka whine or wail)—A more intense, sustained meow. A cry for help when trapped, lost, nauseous, or confused (see pages 180–181). Also used to drive a threat away.

Chirrup (aka chirp or trill)—A soft, short, high-pitched rolled "R" or "prrp" sound with a rising inflection. It's used for locating and greeting other cats, by mothers to find their kittens, when greeting familiar humans, or requesting something desirable.

Chatter (aka chitter or twitter)—A mix of restrained excitement and frustration when desiring unattainable prey (see pages 68–69).

Caterwaul—A long, drawn-out female mating call that sounds like they're in pain. A tomcat's mating call is a distinctive "mowl."

Repulsion tactics

Hiss—An open-mouthed, forceful expulsion of air to deter an approaching threat, or when caught unawares.

Spit—An abrupt pop of expelled air and saliva, often coupled with an equally intimidatory ground-slapping paw.

Growl—A sustained, menacing, and throaty low-pitched grumbling, indicating increasing discontent.

Snarl—An ominous, toothy growl, with a more open mouth and slightly raised lip, flashing their weapons.

Shriek—A sudden loud, harsh outcry or screech during extreme conflict or pain, such as females during mating, or a stepped-on tail: "meowch!"

How kittenhood shapes a cat

Every cat has their own thoughts, feelings, and behavior patterns that shape their personality, partly predetermined by genetics but also affected by their surroundings and experiences. Understanding the impact of nature and nurture means we can ensure that kittens get the right nutrition, positive interactions with humans, and vital life skills that set them up for becoming happy, well-rounded cats.

Sociability tendencies

Whether a cat is bold or timid is influenced by their father's DNA—wary dads make wary kittens. Stress and malnutrition during pregnancy not only stunt kittens' physical growth, but also their psychological development. This can reduce their tolerance of other cats and create more fearful and "grumpy" adult cats (see pages 102–103).

By the third week of life, a kitten's senses and stress hormones are up and running, so every experience shapes their future outlook and behavior. Born with a default wildcat setting, they don't instinctively appreciate human contact—the onus is on us to gain their trust or they'll go feral. Cats must learn to be comfortable around humans through regular positive encounters (gentle handling and play) within the first eight weeks, when their brain is at its most receptive. The window of opportunity for socialization—the most critical and influential period in a kitten's life for their future friendliness—is before you even adopt them. However, all is not lost; cats are not fully mature in a behavioral sense until they're between two and four years old, so there's still plenty to learn.

Eating and hunting tendencies

Many a cat's fussiness is because they grew accustomed to their mother's diet in the womb and in their early weeks of life, as it would have imparted a smell and taste to the amniotic fluid they floated in for 63 days and the milk they suckled on thereafter.

Positive and varied experiences of food in the first six months of life set up a more adventurous palate later on. It's best to expose kittens to plenty of different tastes and textures during that time. Presenting any new foods alongside foods they already like, or having mom nearby, helps acceptance.

All cats have a hunting instinct, but if mom is a prolific hunter, the chances are her kittens will be, too, right down to the type of prey she caught.

SURVIVAL GUIDE

A new cat or kitten

With many cats often reaching the age of 25 or more, it's important to consider your life plan before taking the plunge and choosing a cat or kitten. Here are some points to consider.

1

Best beginnings

Choose a reputable rescue center and avoid online ads and pet stores. Research breeds and ask questions before visiting. Keep your wits about you and trust your instincts.

2

Is age important?

Kittens (from eight weeks), teenage, adult, and senior cats all have different needs. Also factor in that levels of commitment and expense tend to increase with age.

3

***Purr*fect match**

If you'd like a pedigreed cat, match temperament and coat to your lifestyle (see pages 22–25), but don't overlook the well-rounded mixed breed. Both sexes are similar once neutered.

4

Double trouble?

Adult cats don't need a feline friend—just plenty of time with you, the right habitat (see pages 46–47), routine, and a good vet (see pages 152–153). When it comes to kittens, though, happiness is more likely with two.

5

Smooth the way

When you first bring a cat or kitten home, make the transition to their new life gradual (see pages 126–127). Stick to the same brand of food and litter for at least a few weeks and bring familiar scents with them.

6

Keep it positive

Encourage safe, controlled exploration of anything unknown, potentially scary, or frustrating—people, objects, or situations. Affirm calm, confident behavior with familiar treats, toys, and, if they approve, petting.

Cat breeds

The friendly, well-rounded moggy is the result of thousands of years of natural wildcat mating. Pedigreed cats gained popularity between the late 19th century and 1960s—today there are more than 70 breeds. Looks have always taken center stage, but some breeds have inherited tendencies for certain behaviors. Regardless of heritage or appearance, all cats have the same instincts (see pages 10–11).

Breed types

Despite such a variety of breeds, all pet cats share extremely similar DNA to their common wildcat ancestor. We've bred them mainly for their looks, with traditional "natural" pedigree breeds originating from geographical regions, such as the Maine Coon Cat and Burmese. Persian and Siamese cats look very different from their forerunners, as our tastes have changed.

New breeds have been created by humans in several ways:

- Mating traditional "natural" breeds together to create a recognized mixed breed—Tonkinese, Burmilla, Ocicat
- Breeding domestic cats with exotic wild jungle cats or servals to create hybrids—Bengal, Savannah
- Producing offspring from cats with genetic abnormalities, whose unusual looks have struck our fancy—Rexes, Sphynx, Scottish Fold

Some breeds are known for their temperament traits (see opposite), but every cat is unique – a product of their life experiences – so there are no guarantees, especially with controversial exotic hybrid cats.

Bred to socialize
Social breeds, such as the Maine Coon Cat, make great companions.

Breed traits

The movers

All cats need to exercise and explore, but some are particularly athletic and agile, and love nothing more than tearing around at great speed, jumping, climbing, fetching, and even swimming. You can make a small indoor area work if you plan it carefully and utilize the whole 3D space (see pages 46–47), maybe with safe outdoor access (see pages 64–65) and plenty of stimulating playtime (see pages 182–183).

Popular breeds include: Moggy; Burmese; Siamese; Bengal; Abyssinian; Rex breeds (playful) such as Cornish, Devon, and Selkirk; Egyptian Mau (fastest runners); Turkish Van (swimmers)—but not Persians (least active)

The thinkers

Cats aren't calculating, they're curious and highly intelligent explorers. Mental stimulation is extremely important for all cats, but certain breeds are more easily bored and can become frustrated, anxious, and mislabeled as destructive mischief-makers, busting into cabinets and stealing food when under-stimulated. It's in everyone's best interest to keep them entertained (see pages 138–139 and 182–183).

Popular breeds include: Moggy, Siamese, Bengal, Burmese, Abyssinian

The talkers

Any cat can be garrulous with their favorite humans, but certain breeds are more likely to express their love, anxiety, hunger, you name it, by projecting their voice. Meow to that!

Popular chatty breeds include: Siamese, Bengal, Burmese, Abyssinian
Popular quiet breeds include: Persian, Maine Coon Cat

The socializers

Some cats are naturally more social and needy than others (see pages 148–149). They thrive on human interaction, actively seek attention, and become anxious if left alone or if made to share with other cats. If you plan to grow either your human or feline family, it's helpful to choose a breed that's more likely to be cool with that.

Popular breeds for families include: Moggy, Burmese, Birman, Maine Coon Cat, Ragdoll, Rexes, Russian Blue, Siamese
Breeds that can be trickier with other cats include: Abyssinian, Bengal, Siamese, Korat

High-maintenance cats

Always research the time commitment, patience, and expense associated with a breed. Some things might seem obvious, such as longhaired cats shedding fur and needing regular brushing and professional grooming. Others are less apparent, such as hairless breeds being greasy and requiring their own skincare and pedicure routines.

Popular longhaired breeds include: Moggy, Persian, Maine Coon Cat, Birman, Ragdoll, Forest Cats
Popular hairless or sparse and wavy-haired breeds include: Sphynx (hairless), Rexes (reduced fur)

Continued 

Breed features or flaws?

When closely related cats are bred, or random genetic mutations occur in their DNA, their outward appearance can change dramatically. Some features are harmless, while others cause serious health issues, hamper a cat's ability to communicate effectively, and negatively affect a cat's quality of life, either now or in the future.

Look beyond cute or comical appearances. Instead, think about the impact that deliberately passing on abnormalities that are breed norms has on a cat's ability to interact with the world without difficulty, frustration, pain, or illness. Are their "appealing" features actually life-limiting flaws?

Flattened face (brachycephalic)

A shortened nose and jaw means less space to breathe, riskier anesthetics, dental problems, skin fold dermatitis, and difficulty giving birth. Tear ducts are squashed and eyeballs bulge, predisposing them to painful ulceration. No wonder these cats look grumpy!
Examples: Persian, Exotic Shorthairs

Low tail, no tail

The tail is part of a cat's spine and expresses their mood (see page 27). Some breed norms cause restricted (low) tail movement, while others have no tail. This increases the risk of deformity, chronic nerve pain, constipation, and incontinence.
Examples: Low tail—Persian, Scottish Fold; no tail—Manx, Japanese Bobtail

Skin and coat

Some breeds have thinner, curled, or crimped coats, or even no fur at all. This can interfere with normal thermoregulation and self-grooming, and it also increases skin disease,

Breed pros and cons
Always research a breed's weaknesses as well as its strengths. Devon Rexes carry genes for kidney, muscle, and joint problems. They are also prone to problems with their skin and coat.

trauma, and sun damage. Whiskers are often shorter and more brittle, making it trickier to navigate surroundings and interact with prey and toys. Hairless cats are more high-maintenance than a naturally furred moggy.
Examples: Sphynx (hairless), Rexes

Unusual ears

Small, flat, or curled ears are harder to keep clean and leave cats looking like they're in a permanent bad mood. The cartilage defect that deforms ears also causes joint degeneration and painful early-onset arthritis.
Examples: Scottish Fold, American Curl

Short legs

Short legs and long bodies restrict the mobility and flexibility that epitomizes a normal, healthy cat. Short bones and defective cartilage impede running, jumping, climbing, and playing, and are painful once the inevitable arthritis kicks in.
Example: Munchkin

Hereditary diseases

Purebred cats and their crosses draw from a small gene pool, which increases the risk of inherited health problems such as diabetes (Burmese), cancers (Siamese), and heart disease (Maine Coon Cat). The most robust, healthy breed is the one we haven't engineered to our taste—the down-to-earth moggy.

Pedigreed kitten checklist

The breeder is:

- ☐ **Registered with** a reputable organization—Cat Fanciers' Association (CFA), The International Cat Association (TICA)—and has proof of health tests for genetic disorders from both parents.
- ☐ **Comfortable answering** your questions, and has kept a record for each kitten showing regular exposure to handling, other pets, travel, and common household sights and sounds.
- ☐ **Able to provide** proof of vet checks, immunizations, and parasite control, and offer interim pet insurance.

The kittens are:

- ☐ **At least** twelve weeks old.
- ☐ **All together**, with their mom, in the breeder's home (not an outbuilding or cage), in clean, comfortable conditions.
- ☐ **Friendly and relaxed** in your presence, not fearful or anxious.
- ☐ **Comfortable being** handled, stroked, and separated from one another and Mom for short periods.
- ☐ **Fed a complete**, balanced kitten food; have toys; and separate areas for eating, sleeping, and toileting.
- ☐ **All bright**, playful, and mobile. Check that the ears, eyes, nose, mouth, bottom, and coat look clean.

The art of cat watching

Once you're familiar with the different types of communication cats typically adopt to let us know their feelings, needs, or desires (see pages 12–13 and 16–17), you can begin to observe your cat in detail. Your new insight into the messages being conveyed will help you ensure your cat's happiness and strengthen your bond.

How to start cat watching

Keep an open mind

It's all too easy to make assumptions or give things labels, but the key to successful cat watching is to keep a cool head, look at the full picture, and consider the cat's behavior in context. Take an objective view and assess what you see before coloring events with an interpretation. For example, if your elderly cat stops greeting you with an upright tail:
Knee-jerk response: "My cat's getting cranky with old age."
Cool-headed assessment: "My cat sleeps a lot, is stiff when they get up, moves more slowly, pees in front of the litter box, and has matted fur—they're overdue a check-up with the vet."

Film your cat

However closely you watch your cat and their interactions, it's easy to miss the nuances of cat chats in real time. If you're serious about developing your cat watching skills, filming them in different scenarios and watching it back in slow motion or with freeze-frame gives you the chance to study the subtle signs in detail.

Keep calm

If we notice what we consider to be bad behavior, without understanding a cat's way of thinking and responding to things, it can be tempting to launch in and try to stop it. But shouting at your cat or squirting them with water will only make them stressed or fearful, which will make matters worse; staying calm and measured will help ease tension and defuse the situation.

Ask your vet for advice

Your theory about your cat's behavior might be correct, but it's always best to check, even if you're convinced it's a minor problem. Even the littlest thing, or perhaps something you've missed, can start to add up to something bigger (see pages 146–147 and 164–165).

Sunny disposition
= positive mood

Overjoyed tip curl
Tail upright with a curled or quivering tip–may progress to wrapping around your legs

Friendly upright tail
Tail held straight up +/- enthusiastic quiver

Neutral relaxed tail
Loose horizontal tail +/- slow and graceful swishing

Low tail
Covert situations (such as stalking prey), anxiety, pain, injury, illness, Persian breed's trait

Tucked-in tail
Tail held close to the body–apprehension, fear, pain, illness

Tip-flicking tail
Tail tip flicking and twitching–agitation

Thrashing tail
Tail wagging or thumping–fear, frustration, rage

Thundery mood
= frustration and anger

Mood gauge

Observing a cat's tail is a way of gauging how much pressure they're under at any given moment. The position it's held in and the type and speed of its motions tell us a great deal about the cat's emotional state and mood.

Telling tails

Purpose: The tail is an extension of the spine and contains sensitive nerve endings. It is especially helpful for agile, athletic cats, who rapidly change direction at speed and need precise balance at a height. It's a strong visual indication of a cat's mood and intention from a safe distance—ideal for wildcats in long savanna grass.

Position: An upright "flagpole" tail conveys confidence and an eagerness to greet you. A low tail suggests the cat is tentative, anxious, or is stalking prey. It's a breed trait in Persians (see pages 24–25), but can also be a sign of spinal disease or trauma.

Movement: Calm, inquisitive cats may loosely swish their tail through the air in slow, graceful, fluid movements. Abrupt, rhythmic movements, such as tip-flicking, side-to-side wagging, and thrashing are escalating signs of agitation. The faster the tail, the greater the tension.

Signs of a happy tail: A chilled-out, happy cat's tail is loosely held away from their body, somewhere between horizontal and upright. When the tail is completely upright and either curled at the tip or quivering, it's likely that the cat is overjoyed to see you (see pages 52–53). If it's fluffed up like a bottlebrush, the cat is trying to look bigger. That signals fear—so back off!

Watch your cat's tail in different situations, but don't focus on it entirely. It's only one part of the whole communication picture, which can change in the blink of an eye.

Continued

Seeing the bigger picture

A key aspect of successful cat watching is to consider the cat's behavior in the context of both their environment and whatever is happening around them at the time. Reading your cat's body language and vocalization signals is only half the story; recognizing the triggers and stressors that they're responding to is the first step toward addressing them, which will ultimately make life more comfortable and enjoyable for your cat and yourself.

Zooming in

Analyzing just one part of a cat's body, such as their ear position or tail motion, is one piece of the puzzle. If you take other clues into account, such as body posture, eyes, whiskers, and vocalizations, then you can better

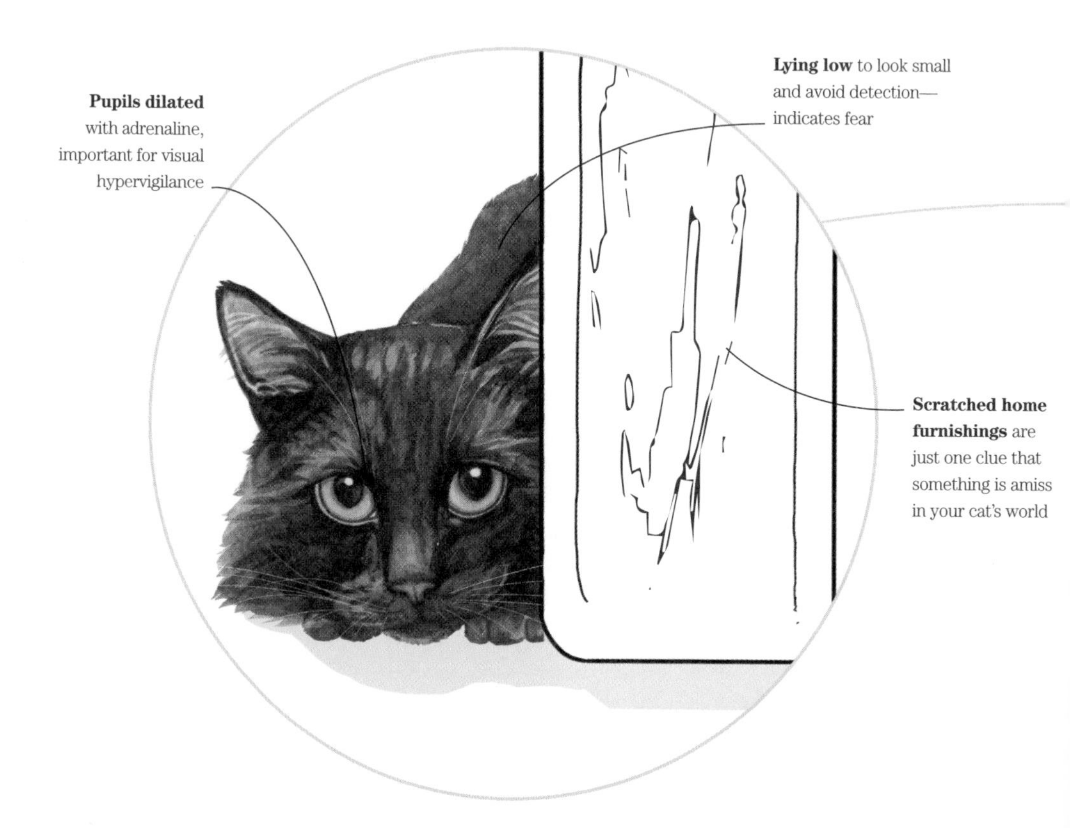

interpret whether your cat's likely feeling anxious, frustrated, threatened, or angry. Similarly, instead of focusing on a behavior, assess the situation as a whole. If your cat is shredding the sofa, resist the urge to mislabel it as bad or unleash your frustration—attempting to discipline a cat is futile and damaging.

Zooming out

Cats are sensitive creatures who soak up tensions in the world around them, leading to a variety of anxiety-driven behaviors. Examining the bigger picture often reveals important details about a cat's surroundings. Hectic environments and unfamiliar or unpleasant sights, sounds, smells, and touch can overload a cat's senses and stress them out. Teach kids to respect cats (see pages 96–97). Create a calmer environment and improve their habitat (see pages 46–47), and offer them lots of quiet, cozy places to escape to and suitable surfaces to scratch and let off steam (see pages 134–135).

A wide-lens view

Pull back and take the focus off your cat's actions. Taking in the full picture is far more revealing.

Trigger stacking

A cat may be able to cope with a single stressful event, but when several occur in succession, anxiety can increase until they reach a breaking point and suddenly withdraw, hide, hiss, or lash out. This gradual buildup of stress is known as trigger stacking.

Even the calmest person can lose their cool at times. If a series of frustrating events sends their day off course—they can't find the keys, get stuck in traffic, and are late for an important meeting—the cumulative effect can send stress levels soaring to such a degree that they're liable to lay into whoever crosses their path. The process is similar for cats, who can be triggered by pain, illness, or anything that makes them feel worried or irritated. The tension builds until it only takes one last provocation—even an innocent pat—to tip them over the threshold into grumpy cat mode (see pages 102–103). At such times, cats need our help to find calmness and our understanding to minimize the effect of their individual triggers or prevent them from arising at all.

Managing a triggered cat:

- **In the throes** of a panic attack, your cat is in "flight, fight, or freeze" mode, so back off and give them space and

How stacking works

Each event raises the cat's tension level, often without you noticing, until they snap.

Stress threshold

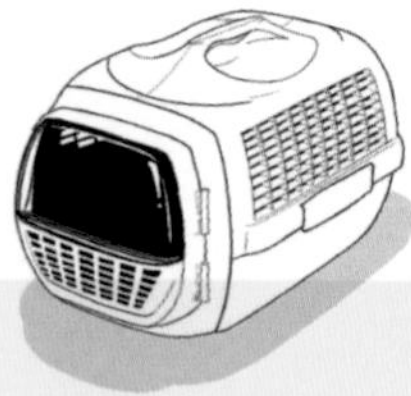

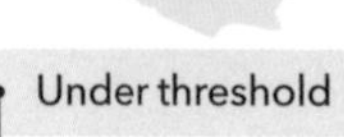

Under threshold

Cat is calm and relaxed

Trigger 1

Cat carrier appears
Triggers negative memories of fear

Trigger 2

Being captured
Loss of control = fear + frustration +/- pain

Trigger 3

Shut in cage/carrier
Can't escape or hide = fear + frustration

time to calm down. Distressed cats like to escape to somewhere low down or high up—preferably dark, quiet, covered, and compact.

- **Don't yell or grab** at a hissing or growling cat, as this will just make you another threat in an already uncomfortable situation. It will increase their stress, and could teach them to fall into a more hostile "armed" response sooner next time.

Preventing trigger stacking:

- **Learn to recognize** the signs of mounting tension through your cat's body language. If you can, let them calm down from the stress of one trigger before encountering another.
- **Help your cat overcome** the things that trigger them by creating positive associations. For instance, if your cat hates their carrier:
- Clean it thoroughly to remove any memorable nasty odors, then leave it open, covered in a familiar-smelling towel, so your cat can investigate it in their own time.
- Create a positive space by adding a cozy blanket and a few treats, and initiating play sessions with a wand toy nearby and, in time, within the carrier itself. Praise their relaxed demeanor calmly and reassuringly.
- Allow your cat time to get used to the carrier while they're under their stress threshold. Giving them choice and control reduces their anxiety and helps them relax, until eventually positive anticipation replaces fear or frustration (see pages 132–133).

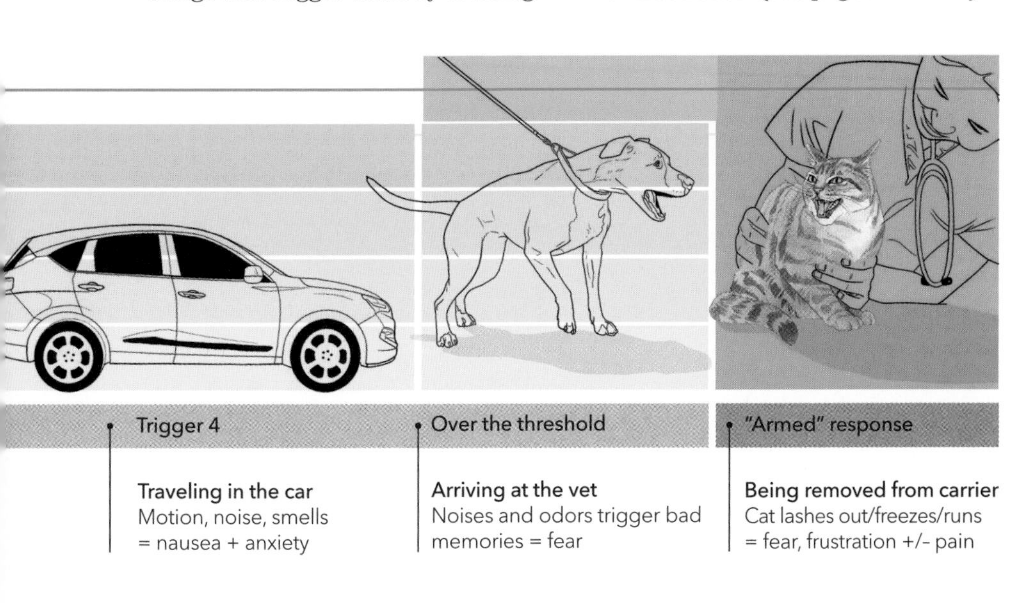

Trigger 4

Traveling in the car
Motion, noise, smells
= nausea + anxiety

Over the threshold

Arriving at the vet
Noises and odors trigger bad memories = fear

"Armed" response

Being removed from carrier
Cat lashes out/freezes/runs
= fear, frustration +/- pain

What's the function?

How a cat behaves in any given situation is determined by a combination of their genetic makeup, life experiences, and an instinctive response to their immediate circumstances. To discover their true motivation, ask yourself: "What's the function for my cat?"

It's tempting to make assumptions about a cat's behavior according to our set of values and outlook on life, but to really understand your cat, you need to put your human perspective to one side and #ThinkLikeACat. The function is key, because cats operate purely from a survival instinct. When a cat behaves in a way that we think is funny, crazy, or cute, there will invariably be a genuine wildcat motivation behind it.

Whenever you are wondering why your cat is acting the way they are, the formula below will help you discover what they are aiming to achieve. Look for clues in your cat's body language, too (see pages 12–13). Next, try to identify the type of behavior (see opposite), which will help you determine what to do next. Be aware that pain and illness can change a cat's behavior, so if in doubt, play it safe and book a vet check.

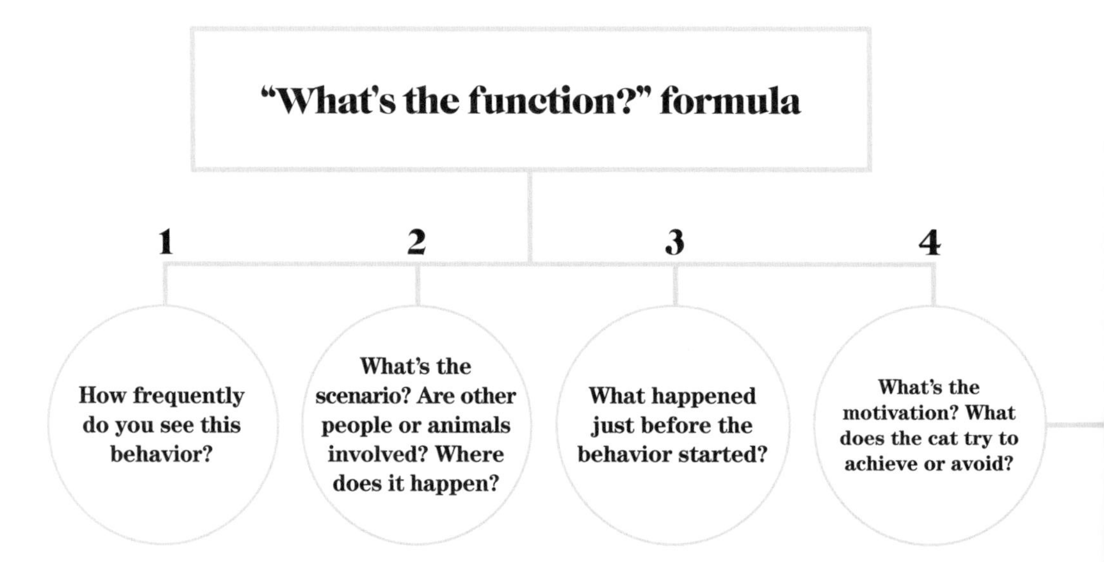

Does the cat gain …

- Security or control?
- An opportunity for normal wildcat behaviors?
- Affection or company?
- Access to something they need or want?
- Distance and time to assess something new?
- Rest, pleasure, or comfort?
- Mental stimulation?

Does the cat avoid …

- Pain or discomfort?
- Frustration or loss?
- Something unfamiliar, unpleasant, or threatening?
- Being cold, wet, or picked up?
- Confrontation or attack?
- Sensory overload?
- Change or novelty?

Behavior types

Understanding which category your cat's behavior falls into will help you figure out what they're thinking and what you should do:

- **Natural behaviors** make a cat a cat, including territorial urges, hunting/playing, grooming, scratching, marking, exploring, jumping, climbing, stretching, hiding, and socializing (on their terms).
- **Learned behaviors** are associations between a trigger and an involuntary emotional or physical response (such as fear, nausea, or taste aversions). They're also the conscious repetition of behaviors that have paid off and the avoidance of ones that haven't. A pay-off or reward could be a treat or unintentional attention—or the behavior might even reward itself (see pages 136–137).
- **Attention-seeking behaviors** aren't dramatic displays of ego for self-validation; they flag that a cat's needs aren't being met. Meowing, jumping up to our level, urinating, begging, and pawing all suggest our attention is needed—something in your cat's world needs addressing.
- **Affiliative behaviors** are feline greetings and gestures used to start or maintain a friendly relationship—mutual licking, nose touching, body rubbing, and co-sleeping with another cat or us.
- **Passive-aggressive behavior** is intimidation without outright physical contact. Common in multi-cat households, it often takes the form of a direct stare, or resting in a strategic location to block an exit/entrance or access to resources such as food, water, litter box, or cat flap.
- **Redirected behavior** is misdirected at something that is not the intended target, such as when a cat sees a rival cat outside and attacks the next person or pet that comes near them.
- **Predatory behaviors** (see pages 94–95) are the drive behind playing, stalking toes under the duvet, or pouncing on your ankles or another cat.
- **Displacement behaviors** are ordinary but oddly timed, and usually signal social discomfort or stress, such as when a cat self-grooms during a standoff with a rival.

CHAPTER ONE

My cat is so cool

Our cats have some seemingly bizarre behaviors that often make us laugh and reach for our phone cameras. Understanding the theory behind these mad moments can reveal a great deal about your cat's well-being—and it's not all funny.

My cat only drinks from a glass

I'm more likely to catch my cat lapping from the glass of water by my bed than from their own water bowl. They also come running for a drink when we turn the kitchen tap on. Why are they being so picky?

What's my cat thinking?

It may seem like your cat thinks they're too fancy to drink from their own bowl, but there is logic behind this behavior. A cat's survival may be jeopardized by drinking contaminated water, so it will seek out running water rather than stagnant pools, and prefers drinking sources that are away from its food or litter box.

Cats also tend to drink when they're relaxed, so a quiet location is preferable. The trouble is, we often put food and water bowls in busy kitchens or utility rooms, where noisy appliances can put a cat off taking a drink. A glass of water in a peaceful bedroom is far more appealing and convenient for a cat to quench their thirst, especially if they've been napping on the bed. And with your scent on the rim, that glass may be just too much to resist.

Glasses and faucets provide a more visible surface to lap from, as well as a raised vantage point, helping your cat stay safely out of the way of other pets or children. This also means they can keep one eye on their surroundings, just like a wildcat at its waterhole.

What should I do?

In the moment:

- **Avoid reprimanding** your cat or shooing them away.
- **Let them finish drinking**—they must be thirsty, after all.

In the longer term:

- **If this is new** behavior, contact your vet; increased thirst could indicate illness (see pages 164–165).
- **Cover your glass** or use a bottle.
- **Try a cat fountain**—the sound and sight of moving water that is always aerated may prove more tempting than stagnant tap water.
- **Experiment** to see whether your cat prefers a narrow or wider bowl, or one made of glass or ceramic rather than plastic or metal. If your cat wears a collar tag, check that it doesn't clank against the bowl.
- **Give your cat options**—introduce multiple drinking stations in quiet spots throughout your cat's territory, with at least one on each level of the home. Try a selection of drinking glasses, bowls, and water fountains.

the function?
Hydration is a survival instinct. Cats prefer drinking from glasses or flowing water as these are often fresher than the water in their bowl and easier to see.
Whiskers back, away from the sides of the glass
Eyes squinting—relaxed, but watching the surroundings
Tongue extended, drawing a column of water into the mouth
Raised vantage point—useful for sussing out threats

My cat zooms around the room

My cat usually lazes around for most of the day, but makes up for it in the evening, haring around the house like they're possessed. Are they just having fun or trying to tell me something?

What's my cat thinking?

Probably, "Clear the way, people, feline oncoming!" Sometimes termed "the zoomies," this rapid switch from placid puss to frenetic fur ball—when a cat, for no apparent reason, suddenly tears around the room as if their tail is on fire—is something many cat lovers will have witnessed. Most cats engage in a mad interlude from time to time, but some zoom more regularly. You may see it after your cat clears their bowels—a phenomenon aptly named "poophoria." If zooming only occurs in this context, it could signal pain, so see your vet. Zooming may also be a clue that your cat needs more stimulation (see pages 182–183).

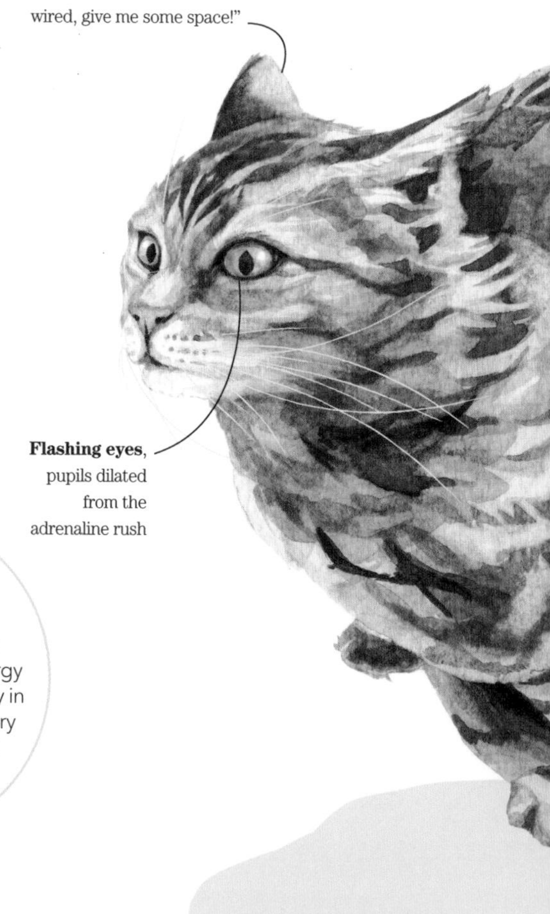

the function?

Zooming is probably an outpouring of surplus energy and/or frustration, especially in cats living a more sedentary lifestyle than they would instinctively choose.

The feline FRAP

Zoologists refer to this sudden burst of liveliness as "frapping"—an acronym from "frenetic random activity period." It usually occurs around dusk and dawn, a cat's natural times for hunting. There's no evidence that cats in the wild do this, but captive big cats, such as tigers and bobcats, do. This supports the theory that frapping is a way of releasing pent-up energy—wild cats are unlikely to have energy to burn when they've been hunting all day.

Swishing tail, used for balance during erratic, high-speed stops and starts

What should I do?

In the moment:

- **Remove obstacles** so your cat can race around at top speed without hurting themself or destroying your prized knick-knacks.
- **Marvel at your hairy hurricane** enjoying their muscle-stretching, heart-pumping workout.
- **Keep your distance**—it can be risky to attempt to interact with your cat while they're so highly aroused, as you might become the focus of some misdirected play and end up a casualty.

In the longer term:

- **Tap into your cat's wildcat** rhythms and give them plenty of physical and mental exercise every day, with items they can climb or scratch and toys to stalk and chase—ideally at times they are typically most active—to ease any feelings of frustration or boredom (see pages 46–47 and 64–65).

Zooming is more common in young cats and those with fewer alternative activities or outlets.

My cat goes crazy for catnip

One of my cats loses all sense of dignity when the catnip comes out. They roll and rub themselves on the floor, drooling and looking as high as a kite, while my other cat remains totally unimpressed.

What's my cat thinking?

Catnip (*Nepeta cataria*) is an herb that releases a potent but harmless oil (nepetalactone) into the air. When inhaled by a cat, it stimulates certain pathways in their brain, and although we don't know exactly what the cat is thinking and feeling, the response definitely seems to be pleasurable.

A cat's response to catnip is an inherited trait, and not all cats carry the "catnip gene." Lions and leopards can yield to the herb's seductive charms, while tigers and domestic kittens seem less affected.

Reactions to catnip vary. Some cats appear to be completely unaffected and uninterested, while others just look very chilled out, although most display a combination of play and sexual behaviors. The most common response is shameless rolling, rubbing, licking, drooling, vocalization, and back-foot kicking, followed by flaking out and purring. This euphoria wears off within ten minutes and leaves cats immune to its effects for around half an hour.

What should I do?

In the moment:

- **Watch how your cat** reacts and interacts after indulging in catnip—do they seem super-relaxed, utterly indifferent, or in raptures?
- **Beware of an intoxicated** cat's sharp claws and teeth, as some get so aroused and overstimulated they may bite or bunny kick your hand.

the function?

Although we don't know why some cats have retained the gene that produces such a reaction, catnip seems to enrich the lives of many cats.

Back legs bunny kicking, to knock out and subdue "prey"

In the longer term:

- **Grow your own catnip** or its less potent cousin catmint (*Nepeta* x *faassenii*), which has pretty purple flowers that attract butterflies and bees. Other plants that have a similar effect include Tatarian honeysuckle (*Lonicera tatarica*), silver vine (*Actinidia polygama*), and valerian (*Valeriana officinalis*).
- **Keep a stash** of dry catnip to rejuvenate your cat's toys and pique their interest in new scratching posts or pads. It's safe and non-addictive, although eating it in large quantities may cause drowsiness or an upset tummy.

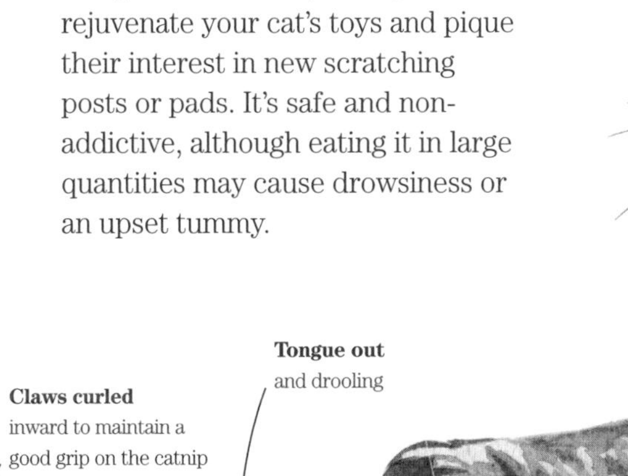

My cat thinks they're a cow

My cat loves tucking into nice, juicy... GRASS! They're also partial to leaves and have even chomped on my orchids. Is plant fiber good for cats? #CatNotCattle

What's my cat thinking?

Like it or not, some cats are grazers. You might not have caught them in the act, but you've probably trodden in the evidence once it's resurfaced on your rug. The jury's out on exactly why meat-eaters consume plant-based fiber, but many wild carnivores munch on vegetation, too. This is likely to be driven by a functional purpose, such as providing extra moisture or nutrients, repopulating the gut flora with soil microbes, or detoxing the gut of poison, excess fur, or parasites. Provided the plant isn't toxic (see pages 64–65), sharp, or laced with pesticides, it's probably not harmful, but it's preferable to supply cat grass in a pot instead.

Lethal lilies

Lilies, but particularly the *Lilium* and *Hemerocallis* species, contain a potent toxin that destroys cats' kidneys. Whether a cat chews the petals, stems, stamens, or leaves, brushes past and licks the pollen off their fur, or drinks the vase water, the consequences can be deadly.

What should I do?

- **Indoor dwellers need** access to cat grass, especially if they're bored or curious. Enrich their lives with ready-made pots, or grow your own by germinating grain seeds, such as wheat, oats, barley, or rye, on your windowsill. Avoid kitchen herbs, as some are toxic if chewed (see pages 58–59).
- **Ensure all vegetation** that comes into your home is safe—spring bulbs, Christmas trees, lilies, and poinsettias are toxic and irritants. Let loved ones know which flowers and houseplants are safe to send you. If in doubt, leave them out—it's just not worth the risk.
- **Be wary of pesticides** or fertilizers if you have a feline lawn-mower.
- **Avoid planting barbed** ornamental grasses—they're literally a pain in the neck, as they can get stuck at the back of a cat's throat and nose when they're throwing them back up onto your carpet.
- **Swap vases and plants** in pots for hanging baskets and terrariums or faux greenery.

Roses, sunflowers, gerberas, violets, and orchids tend to be free from toxins, although prickly leaves and stems should be removed or kept out of reach.

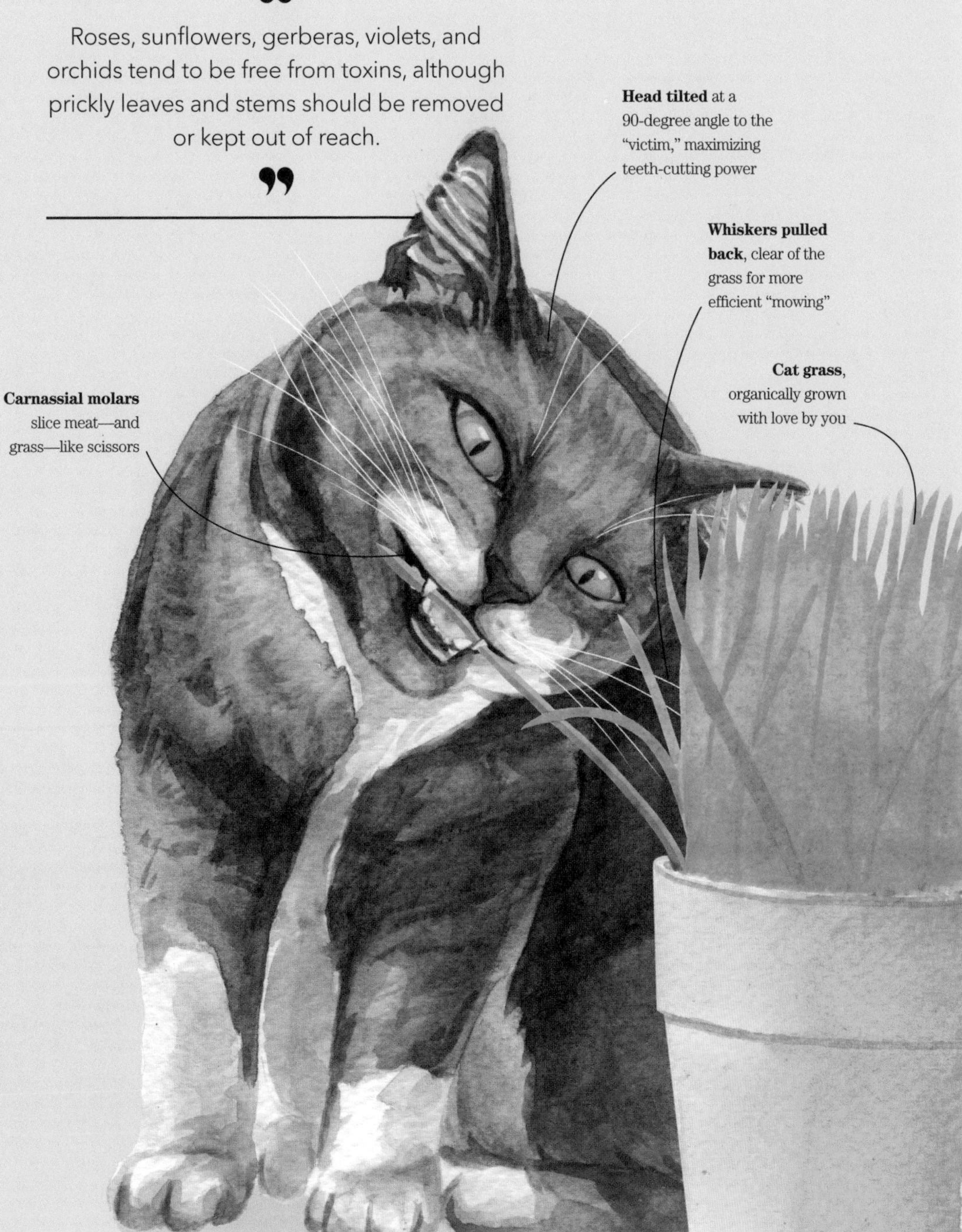

My cat looks down on me—literally!

My cat perches in the highest place in the room, whether it's the curtain rod or even the top of the door! It looks so uncomfortable—why do they do that?

What's my cat thinking?

Cats just love getting up high. It allows them to keep an eye on a large area of their territory and gives them a sense of control. By perching near you but beyond your reach, your cat is interacting with you on their own terms—they're feeling sociable, but not *that* sociable. If another pet has claimed the floor level as theirs, a cat will often look upward to find a safe space to claim as their own.

What should I do?

In the moment:

- **If there's no danger** of them getting jammed in the door or bringing the curtains down, leave them be. When their precarious perch becomes too uncomfortable, they'll hop down. If you do need to get them down, do so calmly, with your safety in mind—use a stepladder, don't balance on the sofa to make a wild grab.

In the longer term:

- **Climbing is a cat thing**—you can't stop it, and nor should you need to. Deal with risky climbing by offering perches that are safer but just as much fun. Try rearranging furniture to make a route from the ground to a higher level. You can buy a cat tree or hammock, but if you're keen on DIY, why not "catify" your space by building multi-level shelves or your own high-rise cat condo?
- **If your cat is out of reach** more than usual, they may feel unsafe on the ground, so offer secure hiding places at all levels. Could they be feeling hassled by another cat (see pages 156–157), a dog, or a child?

Climbing the walls

All but the most elderly or infirm cats enjoy a climb, but agile breeds such as Siamese, Orientals, Abyssinians, and Bengals tend to be the most athletic, with an impressive head for heights. As many owners know, Bengals especially can leap a surprisingly long distance.

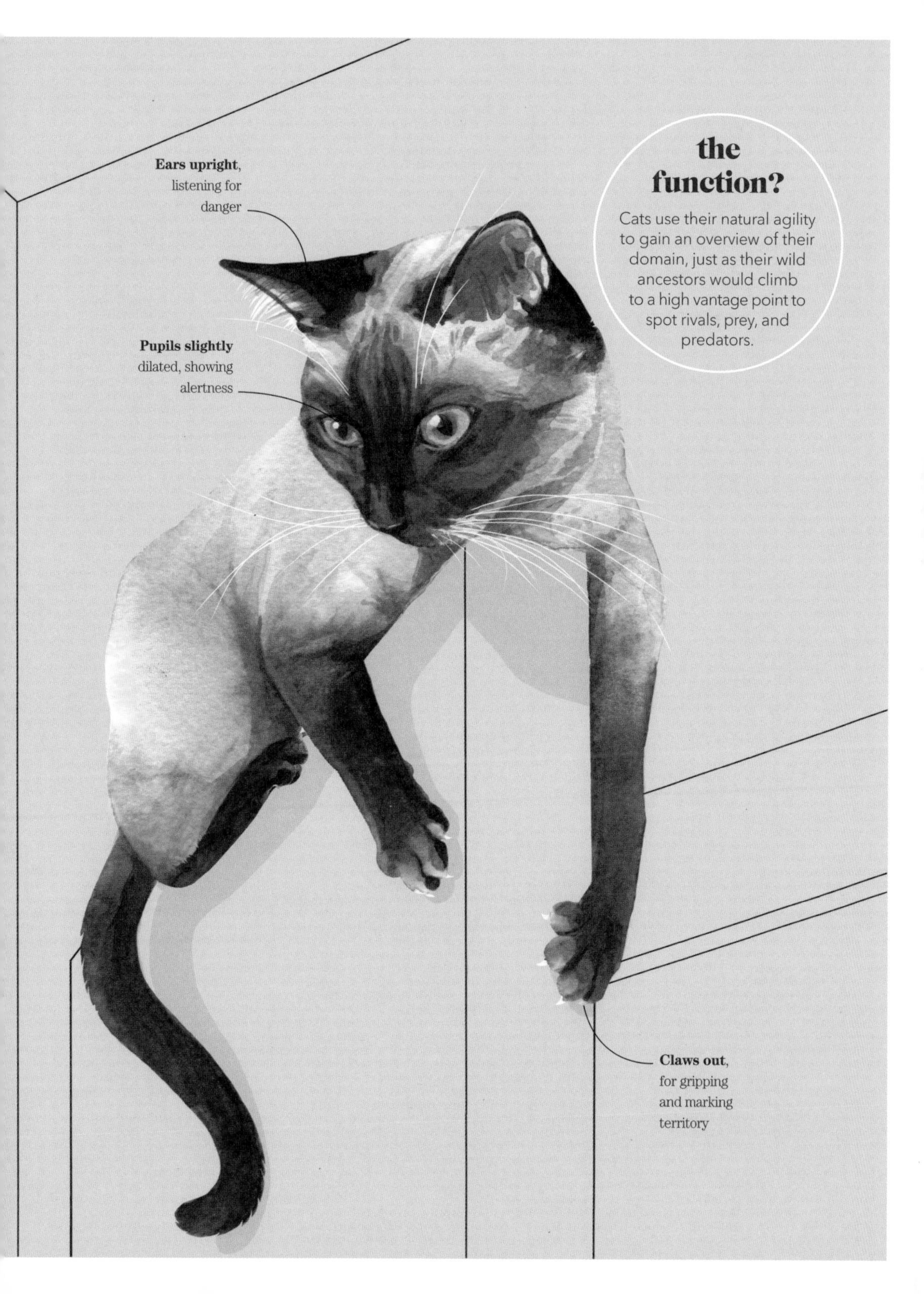

the function?
Cats use their natural agility to gain an overview of their domain, just as their wild ancestors would climb to a high vantage point to spot rivals, prey, and predators.
Ears upright, listening for danger
Pupils slightly dilated, showing alertness
Claws out, for gripping and marking territory

SURVIVAL GUIDE

The perfect cat habitat

Your home is your cat's territory and must provide all they need. It must be a comfortable haven that offers a sense of control and choice, and allows them to be true to the wildcat within them.

1

Make it a zen space

Cats like their territory to be cozy, safe, calm, and quiet, with options for resting and hiding at various levels in different rooms and environments. Let them choose whether they're in the mood for sprawling in the sun or snuggling up tight in dark seclusion.

2

Prevent cabin fever

Cats actively explore their world in 3D, so they need space to jump, climb, run, scratch, and engage with their predatory urges. They also crave new challenges and opportunities to forage and problem solve interspersed with quiet time to rest.

3

A nod to nature

Satisfy their inner wildcat (see pages 10–11), by offering your cat control, choice, and routine. Present vital resources thoughtfully, with separate areas for food, water, resting, and toileting–and with plenty to go around in multi-cat households.

4

Avoid sensory overload

Artificial fragrances, lighting, and noisy tech can overwhelm a cat's super-powered senses. Providing whatever you can in the way of fresh air and green space will be beneficial to your cat's well-being (see pages 64-65).

5

Free movement

Obstructions such as boxes, other pets, children, or visitors can restrict access to a cat's resources or entry/exit routes. Keep interior doors open and make sure mobility issues don't prevent your cat from getting where they want to go.

My cats love home deliveries

When my latest gadget arrives, my cats get more excited than I do. The empty box and inner packaging soon become their new favorite hide-and-seek game. What's their wrap-ture all about?

What's my cat thinking?

Natural curiosity will prompt a cat to investigate anything unfamiliar in their territory, and a cardboard box provides plenty of exciting textures and sounds to explore. They can nestle down in the inner packaging in the same way a wildcat would in a leafy hideaway.

The dark, confined space inside a box offers a nervous cat a refuge from the hustle and bustle of the household, perhaps replicating the feeling of being a kitten cuddled up with their mother and siblings. They also feel protected from "predators."

For a confident, relaxed cat, a box simulates a mini lair. From there, they can covertly survey their surroundings while waiting for "prey" to ambush—hopefully some discarded packaging or a playful fellow feline and not another timid pet or your ankles.

the function?

Any item entering the home from outside will be rich in new smells. A cat's survival instincts drive it to determine whether the object could be a potential threat.

A cardboard box is ideal for predator–prey games, offering a place to hide or stage an ambush

What should I do?

In the moment:

- **Embrace your cat's interest**—the bargain you ordered just became an even greater value for the money!
- **Remove hazards**, such as desiccant pouches or staples. Bear in mind that some cats will chew plastic, sticky tape, or cardboard.
- **Don't let "play"** get out of hand—if ambushing you or another pet becomes the outlet for pent-up instincts and energy, it could result in injury or a fight.

In the longer term:

- **Put cardboard boxes** around the house in high and low places, to provide a choice of play stations or rest spots and to prevent territorial conflict in a multi-cat household.
- **Check out the range** of cardboard scratching pads available.
- **Nurture their inner wildcat** with stimulating fishing-rod toys, catnip, cat grass, puzzle feeders, and kitty TV channels.

Get creative

You (or any willing children) can tap into your inner artist and jazz up a boring box using scissors and paints or marker pens. Just type "cat cardboard box ideas" into any online search engine for some inspiration. Can't be bothered or too busy? There are lots of ready-made options, from castles to cruise ships!

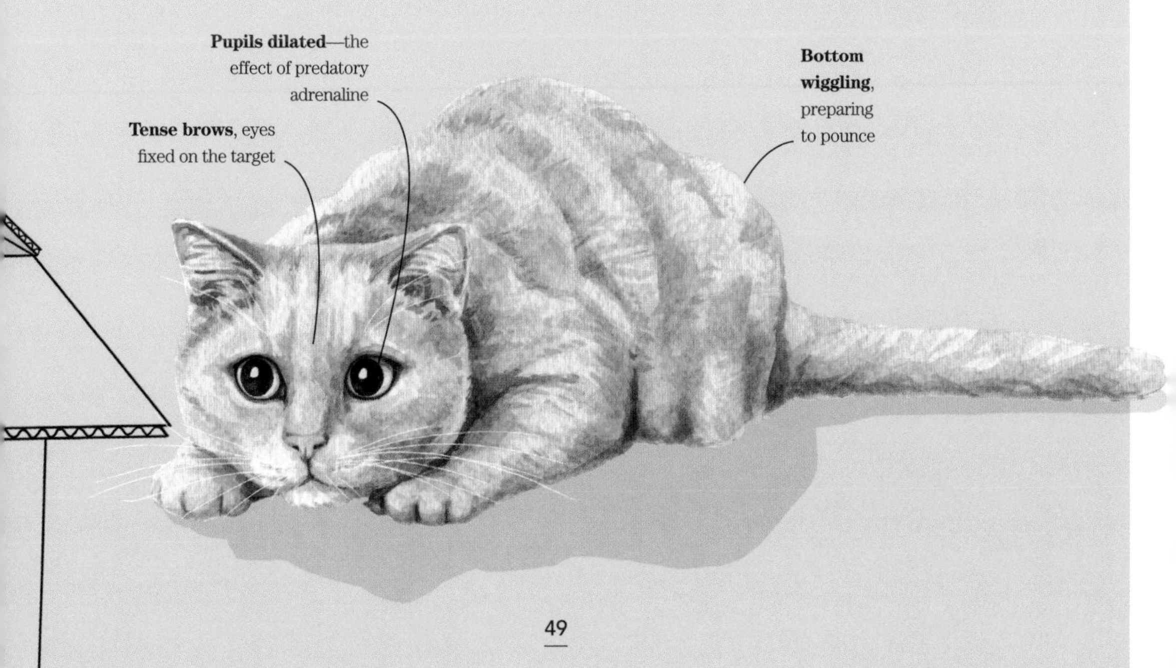

My cat is a clock-watcher

Every day they wake me just before my 6 am alarm and always seem to be up in the windowsill to wait for me to get home after work. How on earth are they so good at telling the time?

What's my cat thinking?

Like humans, cats have a highly developed body clock that dictates sleeping and waking times and affects vital processes such as digestion and temperature regulation. Both internal and external cues keep it in harmony with the earth's natural 24-hour cycle. Cats are expert observers, constantly watching us and gauging what time it is by noticing changes and making associations. Eventually, your cat's body clock incorporates the cues of your schedule—making their stomach rumble just before your alarm goes off.

Cats are very much creatures of habit who thrive on routine, so sudden changes—such as putting the clocks forward or back, or a new work pattern for you—can be stressful and confusing. Thinking ahead can help minimize disruption while they adapt. Of course, cats are always happiest if you fit your life around them!

What's the time?

The "clock" in a cat's brain responds primarily to changes in light conditions, such as sunlight, moonlight, or artificial lighting. Cats also keep track of time by noting our routines, and by detecting the potency of our lingering scent. Other predictable indicators of time, such as birdsong or a neighbor's car starting up, don't escape their notice either.

What should I do?

- **To manage long-term changes**, such as switching to daylight saving time, start the process a week in advance. Each day, move your cat's play and feeding times ten minutes closer to the new schedule.
- **If you'll be home later** than usual or are banking on sleeping in, use an automatic feeder with the timer set at the normal mealtime.
- **The night before** an early start, bring forward the play, eat, and rest routine by up to half an hour.
- **Distract your cat** from reminders that they're home alone—for instance, fit timers to switch on lights at dusk and provide puzzle feeders, cat TV, or calming music.
- **Do not disturb!** Respect your cat's natural daily rhythms and let them sleep when they need to.

the function?

Keeping track of your movements is one of the most functional routines your cat can get into, because you control food, doors, playtime, and fuss.

ADVANCED CAT WATCHING

The cat handshake

Cats aren't antisocial, they're discerning. In fact, they can be socially inquisitive, often soliciting affection from humans they deem worthy. A less-than-ideal greeting (see pages 92–93) is usually down to a human failing to respect a cat's unwritten rules of engagement. A perfect meet and greet looks like this ....

1 The tail-up approach

When a cat approaches with a vertical tail curled at the tip, they're in a good mood; an extra quiver means they're excited. It's the equivalent of a smile. You look approachable so they want to interact with you, but on their terms, not yours–so resist your instinct to go in for a cuddle.

2 The sniff test

Cats suss each other out nose-to-nose, detecting messages of friendliness and familiarity released by scent glands in their faces (see pages 14–15). Mimic this greeting by replicating the shape of a cat's head and nose with your loosely clenched hand, with the middle knuckle slightly extended.

3 Rubbing nicely

Whether it's your feet, shins, or hand, it's an honor to be rubbed against by a cat. Let them guide your hand to where they like to be stroked (see diagram below). If in doubt, stick to the chin and along the cheeks to the base of the ears.

4 The ultimate privilege

A confident cat might push their cheek against your hand and just keep walking for a continuous full-body rub. A really bold cat might even pause for a tail-base scratch, followed up with a single, gentle sweep along the tail. But that's definitely a special gesture and you've really got to understand your cat cues to get that far.

5 Stroking a cat you don't know

Always let a cat make the first approach. Place yourself side-on and sit or squat at their level. Move slowly, calmly, and quietly, and don't make direct eye contact. Stick to the green zones (see right), until you've built trust. Pay attention to the cat's body language, especially their ears and tail (see pages 12–13 and 27). Avoid short, abrupt strokes, which may overstimulate the cat.

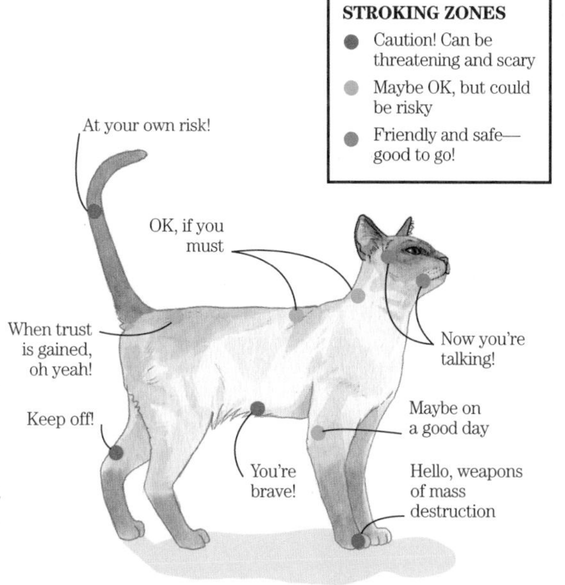

My cat has joined Neighborhood Watch

My cat is permanently stationed at the window, monitoring the movements of every passing dog walker, delivery driver, and fellow feline. Is it just idle curiosity, or should I be worried?

What's my cat thinking?

Window-watching comes naturally to cats: a seat at the window allows them to indulge in two of their favorite pastimes—being nosy and basking in sunshine. The average cat spends five hours a day at the window, and while most are chilled out, probably wondering how tasty that squirrel on the bird feeder would be, some are on high alert, continuously surveying their territory for intruders. An anxious cat is constantly vigilant—dogs and their walkers are potential predators, while neighborhood cats are menacing rivals. The more threats a cat perceives, the more hypervigilant they become. This can lead to chronic anxiety, so worried window-watching of this nature needs to be addressed.

What should I do?

In the moment:

- **Check their body language**—does your cat seem anxious or agitated? (See pages 12–13 and 122–123.) If so, keep your distance and avoid making direct eye contact. The perceived threat of a prowling tom could cause your normally mild-mannered puss to lash out at you.
- **Distract your cat** with a fishing-rod toy that distances you from swiping claws—let the faux fish take the brunt of their frustration.

In the longer term:

- **Out of sight, out of mind**—opaque static-cling window film will help obscure your cat's view of the outside world.
- **Deter feline visitors** by using a humane repellent and cleaning away any lingering urine marks.
- **Keep your cat occupied** and active with puzzles and play that help to use up nervous energy (see pages 138–139 and 182–183).
- **If you have a window** with a less intimidating view, encourage your cat to use it by making it a super-inviting, cozy haven (see pages 56–57 and 68–69).

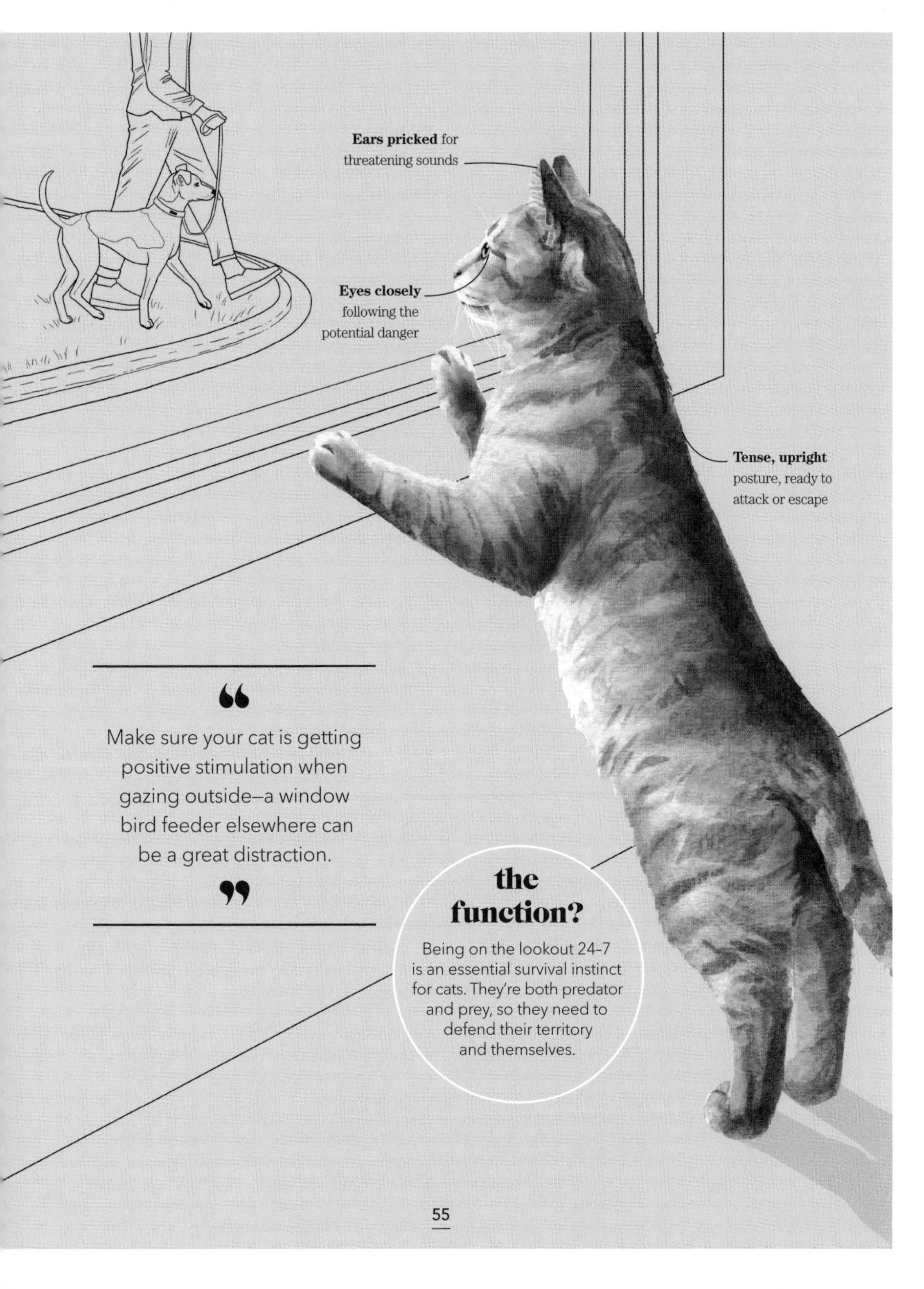

> "Make sure your cat is getting positive stimulation when gazing outside–a window bird feeder elsewhere can be a great distraction."

the function?

Being on the lookout 24-7 is an essential survival instinct for cats. They're both predator and prey, so they need to defend their territory and themselves.

My cat sleeps around

I found the perfect cat bed but my kitty isn't interested. Why is the back of the sofa so much more appealing?

What's my cat thinking?

It might seem as though your cat is channeling their inner Queen of Sheba, but it's humans who have a preoccupation with sleeping in an actual bed. For cats, it's more about a location being safe, dry, and calm enough for them to let their guard down and relax.

A timid, anxious cat might choose a secluded area, such as under a bed, while a confident, sociable cat may prefer a spot in the heart of the home, such as the back of the sofa. A food-orientated feline, however, might favor a shelf close to the kitchen, ever hopeful of a surprise snack. A cat's tendency to rotate favorite sleeping spots is thought to be an evolutionary throwback, to reduce parasite buildup.

Bedfellows or bed hogs?

Your cat will probably view your bed as the ultimate 5-star resting place, with its quiet location, your smell, and cozy bedding to snuggle or sprawl on. You'll need to decide at the outset if they're allowed an "All-Area Access" permit, as it's difficult to impose restrictions later.

What should I do?

In the moment:

- **Let them catnap** in their own way—sleep is a vital time for a battery recharge and body repair.
- **Play cat detective**—why is their chosen spot more appealing than your cozy offering? Could it be the smell, size, shape, temperature, texture, accessibility, or location?

In the longer term:

- **Increase the appeal** of a cat bed by washing away unfamiliar odors and anointing it with pheromones (see pages 14–15).
- **Position a new bed** in a favored location and add a familiar blanket.
- **Ensure the bed** is easy to get into; not too far from food, water, and a scratching post; and away from the litter box and potential "threats."
- **Make the new bed** a positive place, with treats, catnip, or lots of pets.
- **Let your cat choose**—provide different types of beds in various locations: warm beds for heat-seekers, igloo or tunnel beds for hideaway sleepers, or windowsill beds for sunbathers or nosy parkers.

Cats, especially short-haired ones with a slim body type, seek warmth to help them maintain their body temperature. A heating pad slipped into a new bed is a great way to make it more inviting.

Raised vantage point, away from foot traffic

Tail relaxed and loosely hanging

Belly exposed, signaling a state of relaxation

the function?

Having evolved as both predator and prey, cats seek out resting spots that offer safety as well as the opportunity to observe their surroundings.

My cat wants to be a vegetarian

I think they've forgotten they're a carnivore, as they love tucking into strawberries, of all things. They've even stolen broccoli off my plate! Should I buy vegetarian cat food?

What's my cat thinking?

Combine an empty tummy with natural curiosity and you get some unusual food choices. It's unlikely your cat will indulge in a sweet tooth, but they may pick up other tastes they like. Many human foods, such as gravy, salad dressings, and cheese, appeal because they contain salt and fat, like meat does—and of course, something always tastes better when it's stolen!

Food is not all about taste. Cats' mouths are equipped with tools for shearing and slicing through bone, gristle, and tendon, so perhaps your cat is yearning for something that they can literally sink their teeth into.

Nutritional needs

Cats require levels of nutrients only found in animal flesh, which contains vital protein building blocks, such as taurine, and is also rich in essential fats and vitamins A, B, and D. Cats can't digest plant starch as well as dogs and other carnivores, as they have a shorter gut, but pet cats have evolved slightly longer intestines than wildcats, probably by eating a varied diet provided by us.

What should I do?

In the moment:

- **Check that it's wise** and safe for your cat to eat the food they've pilfered. Foods such as milk and cream can cause diarrhea. Other types of food, including chocolate, grapes, and alliums (garlic, onions, chives, and leeks), are toxic.

In the longer term:

- **Is your cat just lacking** other opportunities to explore or play (see pages 64–65 and 182–183)? Would they like some cat grass to munch on (see pages 42–43), or indoor challenges that tap into their natural curiosity, such as food puzzles (see pages 138–139)?
- **Is interaction with you** what they're really craving? Do more things that they enjoy—let them explore open cabinets or cardboard boxes, and get the kitty massager or wand toy out while you're chilling on the sofa.
- **Many cats prefer** a more varied diet than we offer—ask your vet to suggest suitable options.

> There's no evidence that manufactured vegetarian cat food is safe. If your mini-carnivore would eat it, would you be comfortable with that? Food for thought.

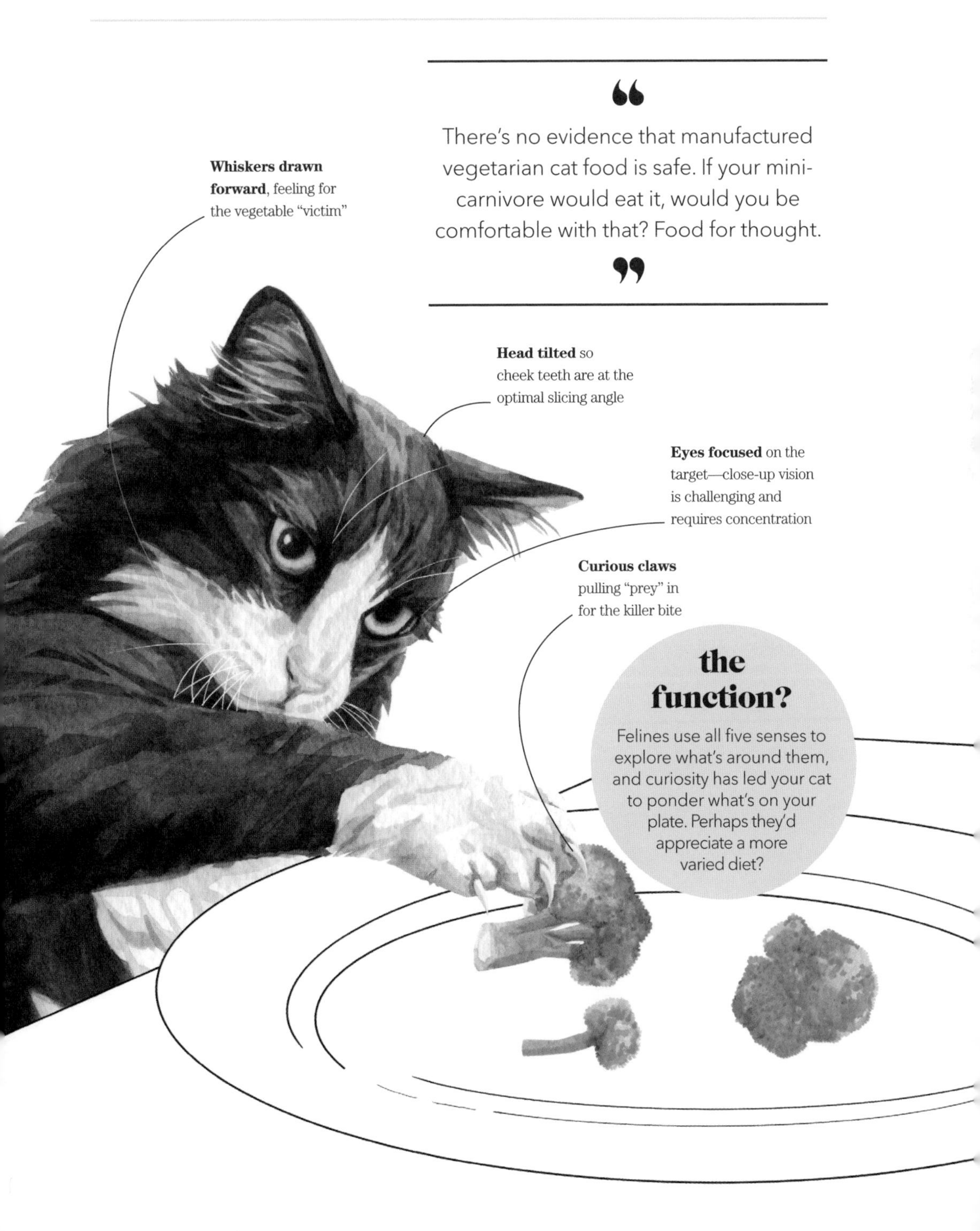

the function?

Felines use all five senses to explore what's around them, and curiosity has led your cat to ponder what's on your plate. Perhaps they'd appreciate a more varied diet?

My cat has gone viral

Their Instagram account gets new followers every day, and the photos and videos I upload get more likes than any selfie I've ever posted. People love my little diva and I love being their purr-sonal assistant.

What's my cat thinking?

The reality is, your cat is completely oblivious to their online fan club and couldn't care less whether the virtual world thinks they're cute, funny, or adorable. Their only concern is whether they're getting love, care, and the right kind of attention from you. Most cats also object to being dressed up, and will probably be thinking, "Get me out of this outfit RIGHT NOW!"

Some cats may seem unfazed when strutting their stuff on the catwalk or striking a pose for the camera, but others will feel irritated or annoyed at having their nap or playtime interrupted and their personal space invaded by a mobile phone taking endless snaps and flashing in their face. Always pay attention to your cat's body language and look for any signs of stress or discomfort (see pages 122–123).

The ethical angle

There's risk attached to fueling trends for cats based on their looks alone, especially when titles like "world's fattest cat," or "amusing" physical features, such as permanent scowls, affect their health or lead to irresponsible breeding. Your cat is not a living toy or a designer accessory, so treat them with respect and don't make fun of them.

What should I do?

- **If your cat struggles**, grumbles, hisses, freezes, or lashes out while you're trying to dress them up or just photograph them, they're not up for it, so respect that and let them go.
- **Avoid extra layers**—most cats have thick double coats and can overheat very quickly if they're dressed up in extra coverings.
- **"Fashion" accessories** can be uncomfortable and dangerous—a diamanté necklace might sound glamorous, and may be okay for a quick photo if your cat is chilled out and happy to comply, but never leave them unsupervised as they could chew it or get it caught on something.
- **Limit how much time** you're spending creating new content for your cat's followers and make sure it doesn't outweigh the quality one-on-one time you're spending with your cat.

“Many cats are true posers or born beauty queens, so focus on celebrating their natural assets and ditch the novelty outfits.”
Headwear irritates sensitive skin and ear whiskers
Pupils dilated with the surge in stress hormones
Costumes are hazardous as they restrict movement and thermoregulation
Plastic claw covers prevent claws from retracting and inhibit natural cat behavior
the function?
Let's be honest, the function is all about you here. Does your cat truly enjoy wearing a cape and tiara, or would they rather play a game or cuddle up on the sofa?

SURVIVAL GUIDE

Exploring the yard

It's not wise to let your cat roam freely, but a securely fenced yard to keep kitty in and intruders out can be a great source of enrichment.

1

Cat fencing

Securely fence all or part of your yard so there are no gaps under or between panels. Add roller bars or an inward-curving section along the top of the fence. Keep your fence well-maintained.

2

Survey your garden

Many plants are poisonous to cats. Check the ASPCA plant database and remove harmful plants. Cats will groom off and swallow what they walk through, so avoid lawn and garden chemicals.

3

Consider a catio

A screened porch or a free-standing enclosed structure are great alternatives to a fenced yard. They can even be accessed from a door or window. Make sure there are spots for shade and for sunning.

4

Preflight check

Make sure your cat is healthy, microchipped, neutered, and up to date with vaccinations and parasite control. Fit them with a quick-release, reflective collar and an ID tag with your phone number.

5

Flaps up

Get a cat flap that will open only for your cat (no squirrels in the kitchen, please!) and that you can lock when you are not around. Train your cat to use the flap, using patience and rewards.

6

Keep them company

Even with a fence, your cat should not go in the yard when you're not home. Predators can swoop down or find the flaw in your fence. An enclosed catio is a better option for home-alone fresh air.

My indoor cat wants a taste of the great outdoors

I live in an apartment, so I can't let my cat outside. But I know the outdoors can be enriching, and I don't want them to get frustrated.

What's my cat thinking?

Cats are natural explorers, so life indoors can get boring and stressful. There are plenty of dangers outdoors, but it also offers variety and a chance to exercise many natural behaviors, including stalking and hunting. Every cat should have a secure and interesting indoor habitat that meets their wildcat needs (see pages 46–47 and 162–163); if possible, they should also be offered safe ways to expand their adventures to the great outdoors (see pages 62–63). If it's safe, you can also let your cat roam your apartment hallway and pretend they're outdoors.

Strolling and rolling

Confident cats can be taught to return on cue, accept a harness, or ride in a pet carrier or stroller. Anxious cats, however, find a changing environment and the inability to escape, hide, or control how they explore too upsetting. Consider your cat's personality, try out anything new indoors first, and monitor your cat's body language for signs of stress (see pages 12–13 and 122–123).

What should I do?

Bring the wilderness inside:

- **Foraging fun**—fill a cardboard box or paper bag with leaves or small branches of cat-safe plants and let your cat explore. Make sure anything you pick up has not been treated with pesticides or other chemicals.
- **Wildlife watching**—set up a bird feeder outside the window. If you have a fire escape, put out a pan of birdseed.
- **Screen nature videos** featuring birds, squirrels, mice, and other small critters—there are many to choose from on the Internet. Make sure the sound is on because that's as exciting as the visuals.
- **Create a cat-safe** indoor garden (see pages 42–43), with cat grass to nibble on, and nontoxic plants such as parlor and areca palms, Boston or sword ferns, and spider plants from which to stalk and ambush faux prey.
- **Make a cat space** on a sunny windowsill—clear off all your stuff and add a kitty cushion. Consider a window-mounted hammock.

Nose detecting nature's scents in the fresh air
China Aster (*Callistephus chinensis*)— nontoxic for cats
Pupils constricted in the bright sunlight
Sensitive pink skin prone to UV sun damage
Ears upright, forward, and alert for unusual outdoor sounds
the function?
Cats are designed to interact with the natural world, and will appreciate it no matter where they experience it.

My cat brings me gory presents

It's like I'm living with a feline Hannibal Lecter and my bedroom's their mini-lair. I love my cat, but I also love other animals. Can my little serial killer be rehabilitated?

the function?

Hunting is a survival instinct. Prey is brought back to the core territory, away from other cats, scavengers, and predators, to be stored for later or eaten in safety.

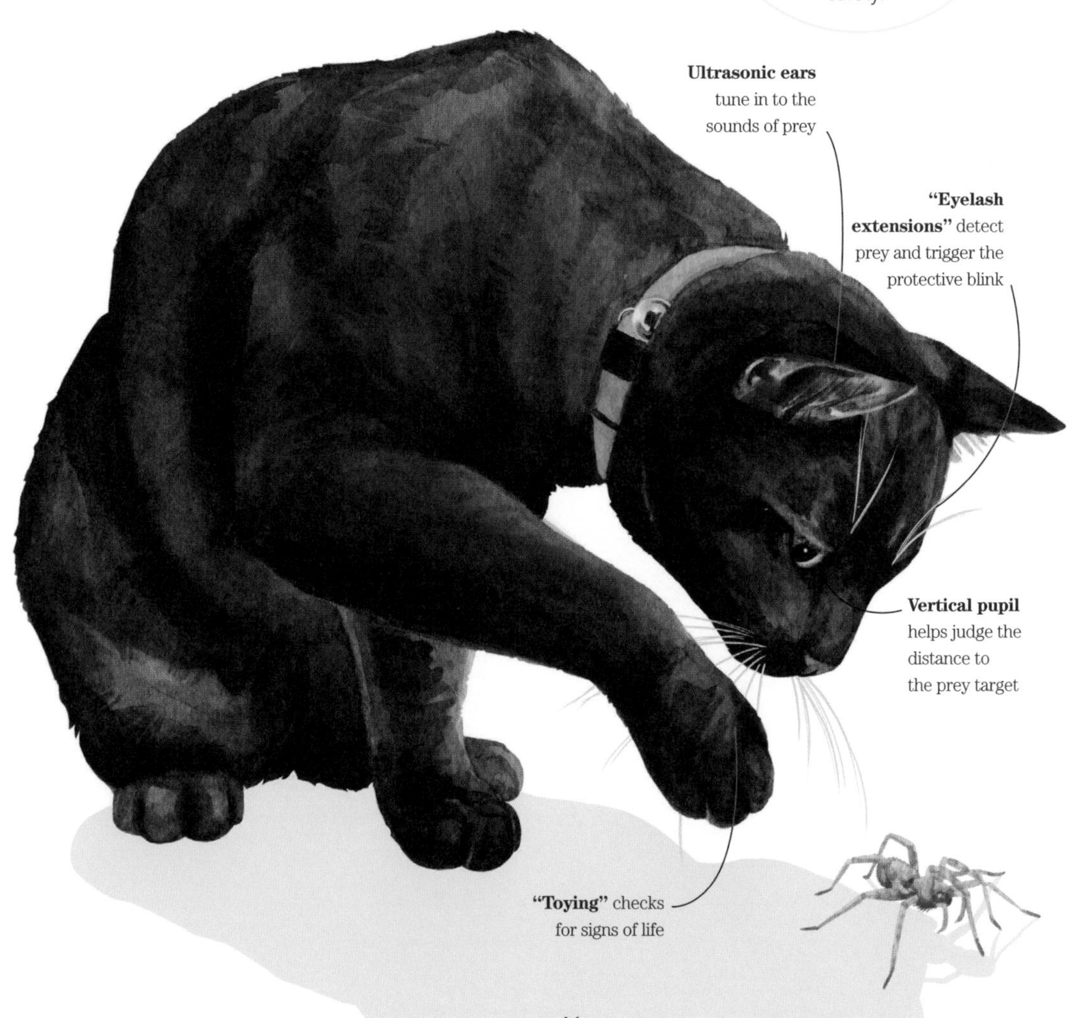

You can't punish cats or stop them from expressing a natural behavior—all you can do is give them an alternative outlet.

What's my cat thinking?

Whether it's an occasional offering or a daily ritual, some cats bring you gruesome or live prey, from mice or birds to spiders and moths. You may interpret this as oversharing hunting joy or a loving gesture, but your cat is simply listening to the call of the wildcat within. If your cat's mother was a prolific hunter, they will have been taught well as a kitten. Whether they actually eat their victims may depend on their preference for the taste of fresh mouse or crunchy fly over the cat food you give them. Either way, hunting is an instinct, and even pampered pedigrees may relish the opportunity if it presents itself.

What should I do?

In the moment:

- **Keep calm**—don't scold, chase, or reward your cat with treats or praise.
- **Letting nature run its course** can be traumatic and messy. If you can, secure prey in a small box, somewhere warm, dark, and quiet, to reduce suffering while you call your vet. Injured prey may survive with the right medication and expert care, but the shock and risk of infection means simple release probably isn't in its best interest.

In the longer term:

- **Make sure your cat's** parasite control is up to date, particularly if they like to consume their catch.
- **Mimic nature** with small, frequent, meals foraged from food toys (see pages 138–139) and plenty of simulated hunt activity, aka play (see pages 70–71 and 182–183).
- **Hamper their chances** of hunting success. Block up the cracks and crevices where critters may be coming into your home. Avoid poisons, which your cat may eat.
- **Remove the bird feeder** if you have been letting your cat spend time in the yard (see pages 64-65).

Cats vs. feathered friends

The need to protect precious native fauna while respecting a cat's natural predatory urges is a controversial moral juggling act. We've invited cats into our homes, but nature's original pest-control doesn't spare endangered species, so it's wise to use curfews and a quick-release collar with a bell to help protect wildlife.

My cat chatters at the birds

Sometimes they make the weirdest noise when they're watching birds and squirrels on the feeder outside. It's like they're trying to communicate with them, willing them to come into reach.

What's my cat thinking?

One of the more amusing noises in a cat's vocal repertoire, chattering, chittering, or twittering is usually reserved for prey-watching. If you look closely, you'll see a rhythmic twitching action of your cat's muzzle and jaws as they vocalize while opening and shutting their mouth in quick succession and chattering their teeth together. Each cat seems to have their own signature sound, which is a variable combination of clicks and short, sharp squeaks. Whether they're on virtual pigeon patrol from a cozy window seat or glaring at an out-of-reach squirrel, the common feature seems to be seeking the unattainable.

Like an adrenaline-pumped racing driver revving their engine at the starting line, a chattering cat is probably experiencing a mix of excitement and anticipation.

What should I do?

In the moment:

- **Enjoy the entertainment**, but be mindful of the frustration that potentially lies beneath this behavior, perhaps because of limited resources indoors for them to explore, climb, stalk, and pounce as much as they'd like to (see pages 46–47 and 64–65). See this as an opportunity to make sure your cat has plenty of healthy stimulation.

In the longer term:

- **If wildlife-watching** is your cat's favorite pastime, embrace it and make a comfy spot with a view for your little twitcher. Window seats for cats come in all shapes and sizes, including hammocks that hook over radiators or attach securely to the glass pane with suction cups, or you could brush up on your DIY skills and create your own.
- **Nurture your cat's wild side** with new toys that mimic their favorite prey—go for feathers for bird-lovers and faux fur for mousers.

Hunting strategy

You'd think that predators who serenade their prey while stalking them would be at risk of giving themselves away, but South American margay cats make noises that mimic the babies of their forest prey (pied tamarin monkeys). Perhaps pet cats are cleverly luring their prey, too? Or it could just be a mix of predatory excitement and frustration.

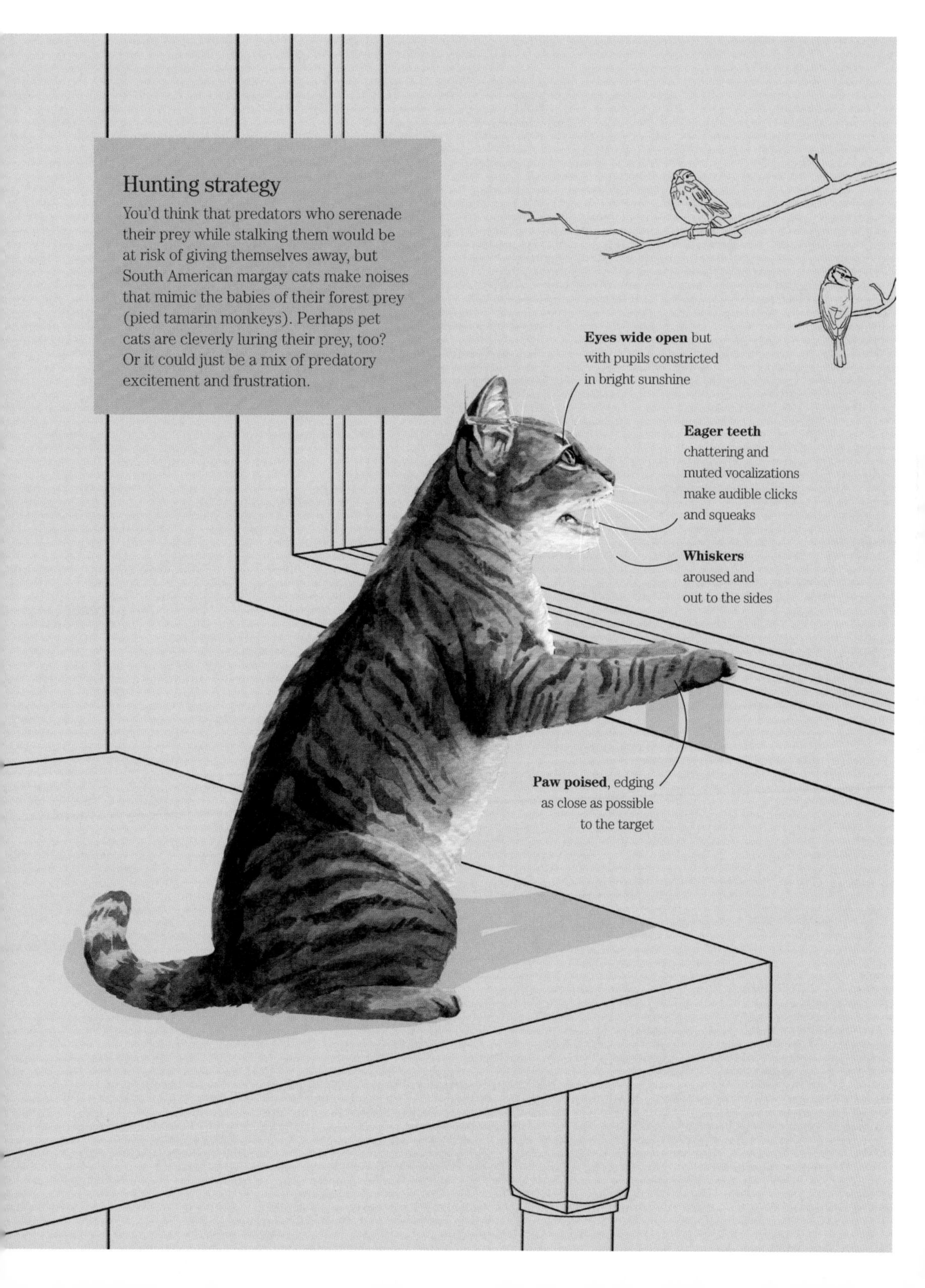

ADVANCED CAT WATCHING

The prey sequence

Predatory behavior is a fundamental part of being a cat, and whether they're prowling for a live target or ambushing faux prey, their tactics will be similar. Understanding the prey sequence is a great way to learn about your cat's natural instincts and helps you create stimulating play scenarios.

1 Search

Cats patrol their territory alone, looking for suitable prey (small rodents, birds, reptiles, or insects). They will sit and wait patiently, their ultrasonic hearing tuning in to high-pitched noises and scurrying. Their eyes scan for clues before they focus their stare on the target's location.

2 Stalk

Cats often sneak up on their prey from the side or rear. With their eyes on the target and legs bent, they slowly and steadily creep forward, tummy to the ground. Things can change quickly, so they may run at times, while keeping their squat posture.

3 Chase

As ambush predators, cats rely on stealth and the element of surprise, rather than pursuing and exhausting their prey. If they run, it's not a long-distance chase, it's a sprint–sometimes with erratic twists and turns, as required, to stop the prey from escaping.

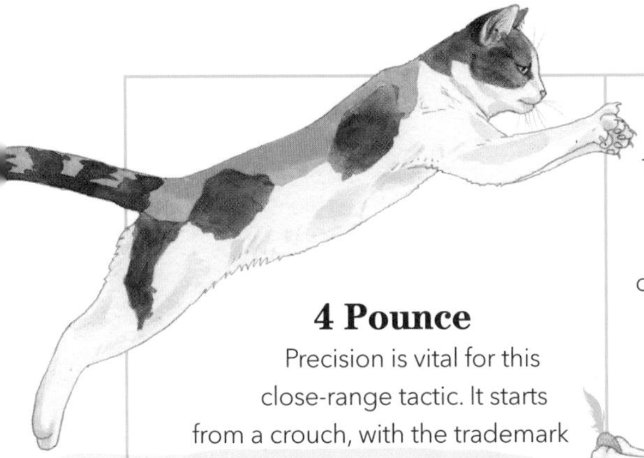

5 Capture and kill

The prey is caught between paws or jaws. With poor close-range vision, cats use their front paw pads and claws to detect movement, while their lips determine the prey's orientation and canine teeth deliver the fatal bite.

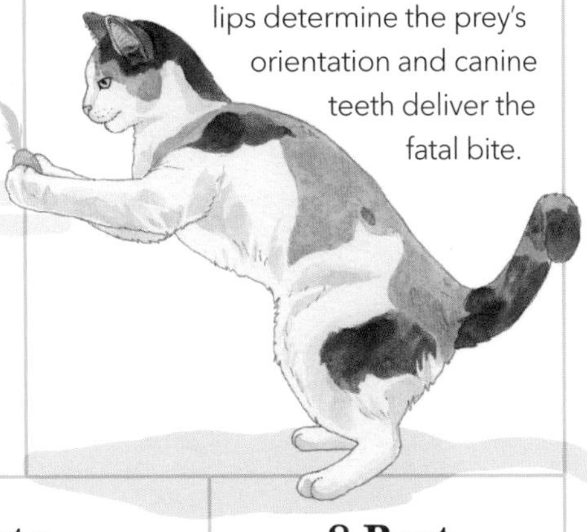

4 Pounce

Precision is vital for this close-range tactic. It starts from a crouch, with the trademark "butt wiggle," before the body is propelled forward and upward. Focus is on the target, with head lowered, eyes fixed, and ears pricked. On landing, whiskers sweep forward to gauge the prey's exact location.

6 Manipulate

Prey often fights back if not killed instantly, so cats bat, toss, and fling small victims and bunny kick larger ones. They may briefly let their target go to "test" for signs of life or to avoid getting injured.

8 Rest

This arduous process is replayed up to 20 times a day in the wild, so recharging is essential. This is the time to digest their kill and wash blood and parasites off their fur, so as not to attract predators or scare off future prey.

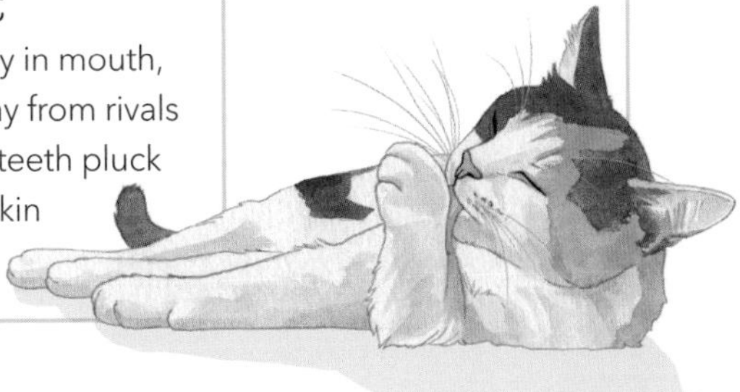

7 Prepare and eat

The hunter will slink to a safe spot, prey in mouth, so they can either tuck in or stash it away from rivals and predators. If they're hungry, front teeth pluck feathers and a barbed tongue strips skin and fur, exposing the flesh.

CHAPTER TWO

My cat and me

Cats don't broadcast their emotions or hang on our every move—they often prefer to just quietly observe from a distance. It doesn't mean they don't dote on our company or enjoy our affections, they just value their own company and need their space.

My cat kneads and drools on me

I love snuggle time with my cat—apart from the fact that they dig their claws into my lap and often leave a puddle of drool. #NotCool

What's my cat thinking?

You've probably seen your cat rhythmically pushing their front paws back and forth, like a baker kneading dough, either on your lap or on a bed or blanket. This indicates they're getting positive vibes—their wild ancestors did it to help them bed down before sleep.

This behavior, or the cutesier description "making biscuits," is also a normal instinct in suckling kittens, to encourage their mother's milk to flow. Some adult cats even suckle on soft items, such as clothes or blankets, when they're kneading (see pages 116–117), while female cats in heat often knead the ground.

The pleasure and relaxation your cat feels during snuggle time with you can also trigger the production of excess saliva (as if preparing to digest mother's milk). It's the opposite of an adrenaline rush—their body goes into "rest and digest" mode, rather than the "fight or flight" survival mode. They relax, their heart-rate slows, and watery saliva is released. Excess nasal secretions are also stimulated in some cats, resulting in a drippy, wet nose.

What should I do?

In the moment:

- **Stay calm**—you don't want your cat to go from charmed to alarmed or they might avoid your lap in future. You can clean up later.
- **Make a mental note** to check that your cat isn't drooling, other than when they're euphorically kneading, purring, and content. Excessive drooling can be a sign of dental disease, nausea, toxins, insect stings, or an obstruction, which all need prompt veterinary attention. Keep an eye on your cat's appetite and eating behavior in particular.

In the longer term:

- **Keep a blanket or cushion nearby** so you can place it on your lap in readiness, before your cat settles down to snuggle—and you can avoid feeling like a human pincushion.
- **If you have an indoor cat**, you could carefully trim their claws (cats who spend time outdoors need theirs for climbing and quick getaways).
- **Keep tissues on hand** to soak up the drool.

Drool, dander, and allergies

People with a cat allergy react to a protein in cat saliva, Fel d 1, which is left on the pet's fur during grooming. When a cat scratches, or is stroked or brushed, protein-rich hair particles (dander) become airborne and can be inhaled. Specially developed cat food neutralizes salivary Fel d 1 while it is chewed.

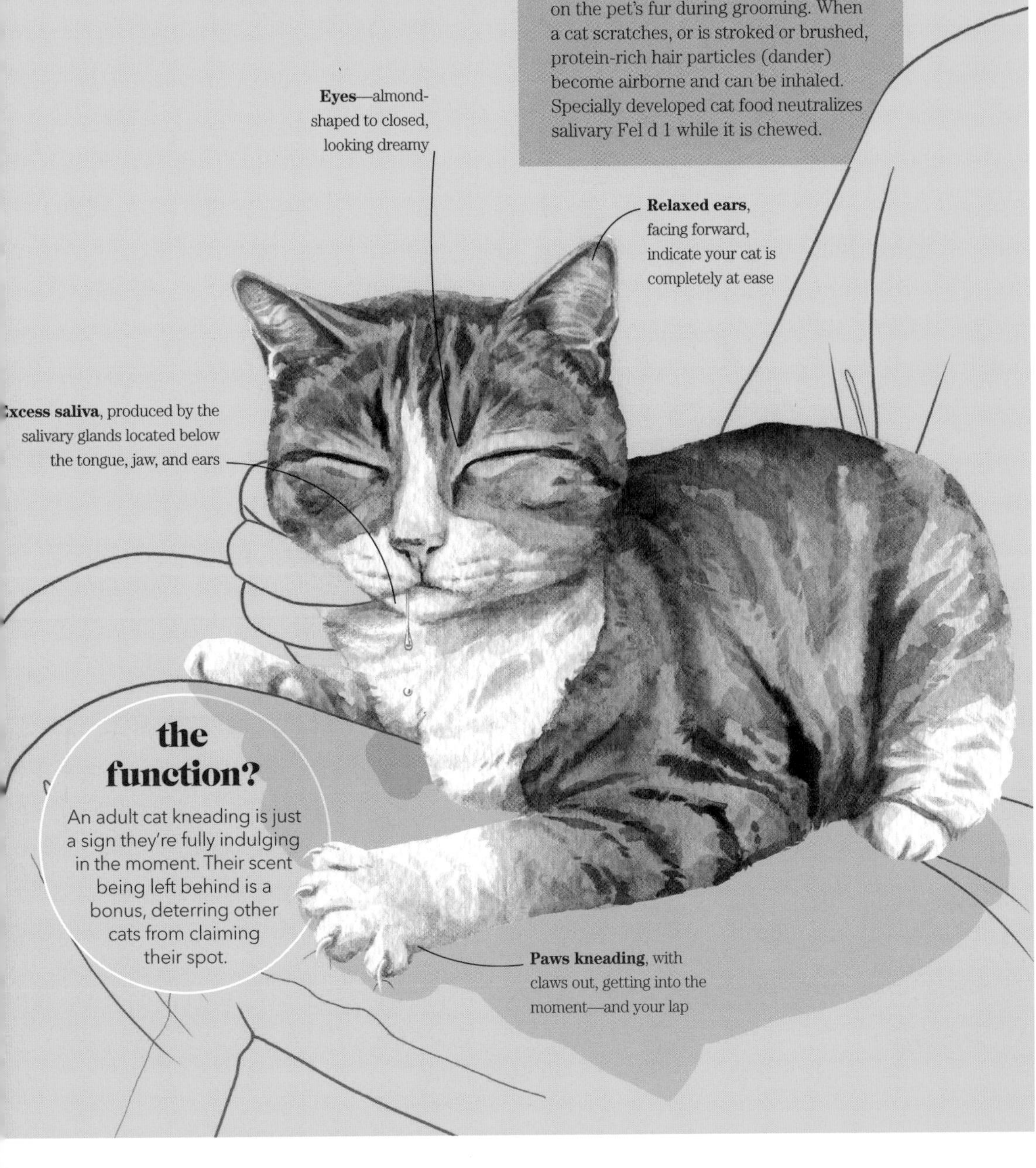

Eyes—almond-shaped to closed, looking dreamy

Relaxed ears, facing forward, indicate your cat is completely at ease

xcess saliva, produced by the salivary glands located below the tongue, jaw, and ears

the function?

An adult cat kneading is just a sign they're fully indulging in the moment. Their scent being left behind is a bonus, deterring other cats from claiming their spot.

Paws kneading, with claws out, getting into the moment—and your lap

My cat thinks I need grooming

Sometimes my cat takes our relationship to a whole new level and starts licking my hair and skin. Are they showing affection, or do they think I need help keeping myself clean?

What's my cat thinking?

This behavior is most likely a sign that your cat is confident you are a trustworthy part of their social group and they just want to enhance your relationship and strengthen the bond between you by sharing the group scent (see pages 14–15). Could they be expecting you to return the favor? Fortunately, a pat is probably enough to keep them sweet.

Being licked by a cat's barbed tongue can be unpleasant (and even painful), which is not surprising when you remember that it's designed to strip skin and meat from the bones of prey. If it's any consolation, this gesture comes from your cat's gentle, peace-making side, and would have been learned from their mother licking them in this way when they were a kitten.

Keeping the peace

Mutual grooming, or allogrooming, may help defuse anxiety and tension between cats in the same social group, as it does in other species. It's often misinterpreted as a sign that cats are on good terms with each other, but, like all friends and families, cats don't always see eye to eye and may lick each other to soothe and avoid a full-blown fight.

What should I do?

In the moment:

- **If you like it**, do nothing—it's harmless and increases your bond.
- **If you don't like it**, avoid scolding or scaring your cat, as this could break the circle of trust.
- **Note any triggers**—what was happening just before your cat decided to give you a feline facial?

In the longer term:

- **Distract and redirect** them with an exciting alternative activity, such as play, whenever you spot those early warning signs that your cat's getting ready to bathe you.
- **If the time and place** suit you, and it's just the rasping tongue that's the issue, initiate a stroking session instead. Pay particular attention to their head, where all those scent glands are (see pages 14–15), and they'll be reassured that your scents are sufficiently entwined.

the function?

It's normal for cats to groom their cat friends—like a feline hug between best buddies. Allogrooming gives both parties an endorphin rush.

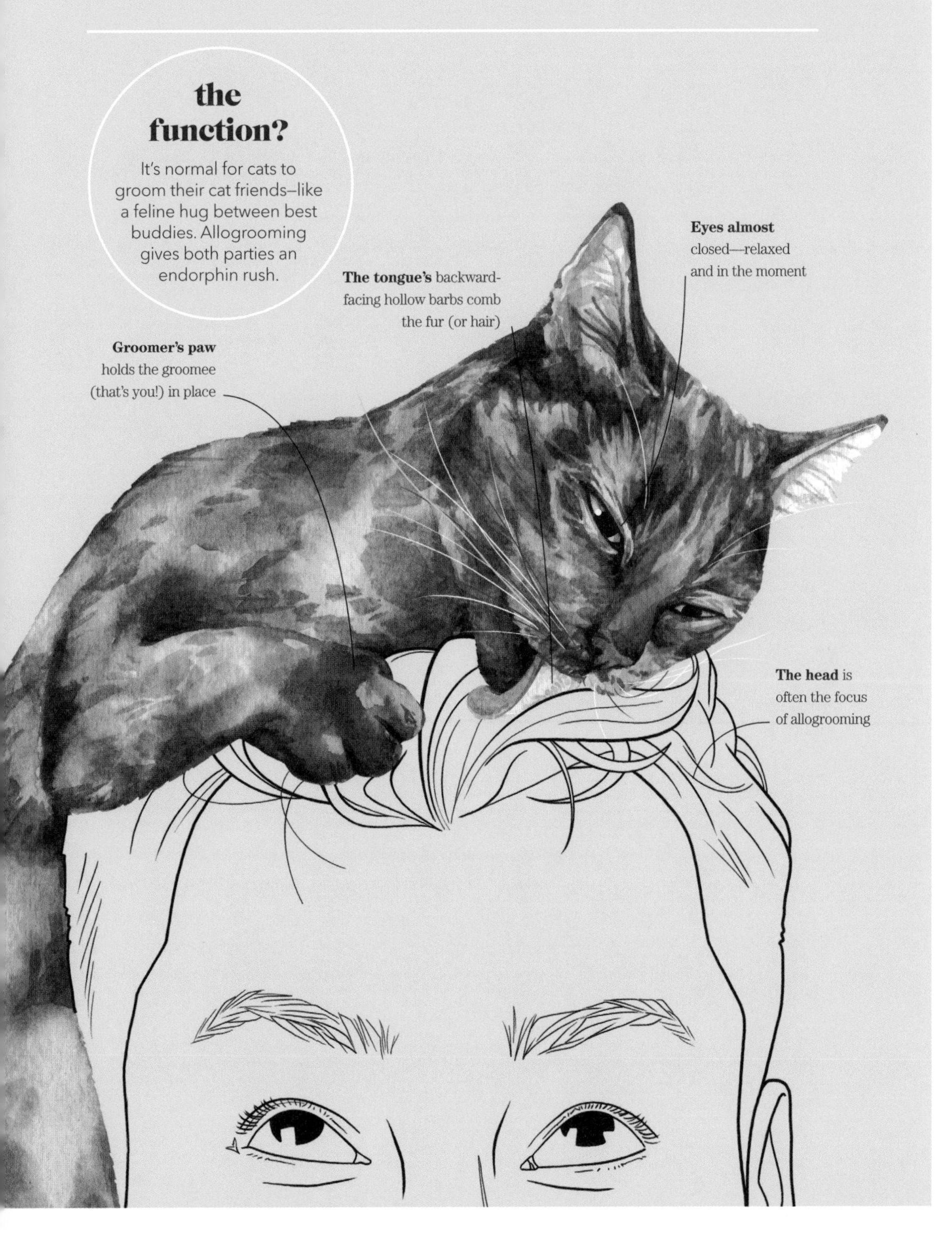

My cat out-stares me

I relish the challenge when my cat wants to have a staring contest—but I always lose and blink first. I guess they might be trying to tell me something important, but how would I know?

What's my cat thinking?

Cats are driven by the wildcat within them to be constantly visually aware and curious about their surroundings. They think nothing of staring at a mousehole for hours. They also use the power of the glare during a standoff with a rival, often accompanied by growling or yowling. It's a battle of wills until one backs down or a fight ensues.

Humans are programmed to find a prolonged, unblinking, and direct stare a tad unsettling, too. Cats are far better at reading our body language than we realize, so maybe they can sense this takes us out of our comfort zone. They will have learned from past experience that if they stare at us for long enough, they'll get our attention.

In many ways you're the center of your cat's world. You control their access to food, water, shelter, litter box, health care, entertainment, and the attention they receive. Perhaps your cat feels their needs aren't being met in some way, or maybe they're just naturally curious about what you're up to and why it doesn't involve them.

What should I do?

In the moment:

- **Notice where you are** and what's happening—this could give you clues as to what your cat is trying to tell you.
- **Read their body language**—if they're scared and hiding, agitated, angry, or gearing up into play/predator mode, avoid staring back and give them space.
- **Check for signs** of stress or health problems (see pages 164–165).
- **If you think** your cat wants food they don't need, break eye contact and provide a distraction, such as a play session (see pages 182–183).
- **If it's affection** they're after, make the most of a good petting session.

In the longer term:

- **Work on reducing** any frustrations or stresses in your cat's life. Provide routine and structure, but allow them variation and control, too.
- **If it's food-seeking** behavior, review what, how, and when you feed them to see if you can better match their needs.

the function?

Staring is a survival instinct that helps cats avoid danger, such as injury from predators or rivals, and it forms a key part of a cat's sit-wait-ambush hunting style.

> "Cats can go longer periods without blinking than humans can, so you're unlikely to win the staring contest."

My cat headbutts me

Every time I settle down to watch TV, my cat pesters me with their trademark move, the headbutt. They've even done it with such force that my hot coffee sloshed all over me and the sofa!

What's my cat thinking?

No prizes for guessing that your cat's trying to interact with you, but what does their ultra-pushy head-nudging actually mean? Bunting, as it's termed, is a friendly gesture that even big cats such as lions use to affirm their bond with each other. With your cat, it probably started as a subtle, cute, and gentle head rub, and progressed with time into an over-enthusiastic headbutt. This is often teamed with a loud purr and, if you're lucky, another affectionate gesture such as the drool, lick, or blink (see pages 74–75, 76–77, and 86–87). At some point, your cat has learned that when they perform this routine, they get a lovely neck and head massage out of it.

Headbutt or headache?

Cats seem to bunt humans with greater enthusiasm than in a cat-to-cat gesture. It could be that they're communicating more forcefully because we're not responding in the way they're hoping we will—just as we'd raise our voice if someone hasn't heard us.

What should I do?

In the moment:

- **Only bunt back** if you want to encourage the behavior—not wise when you're holding a hot drink.
- **If the timing is not right**, break the cycle by not giving your cat what they want at that time and in that context. Move away and ignore them if you have to.

In the longer term:

- **Give your cat** an opportunity for some supercharged bunting at a time more acceptable to you—for instance during a grooming session or in between playtimes.
- **Before you sit down** to relax, play an active game to allow them to vent their pent-up energy. Alternatively, keep them busy with a self-directed activity, such as a catnip toy or a puzzle feeder (see pages 138–139).
- **Teach your cat** the benefits of expressing affection more calmly, by encouraging calm, low-energy rubbing sessions, focusing on the underside of their chin and cheeks, rather than the top of their head.

the
function?
Cats within a social group mix their facial pheromones by rubbing and bunting each other, so a stranger can be easily identified with a sniff (see pages 14–15).
Bunting—top of the head angled for maximum contact
Eyes soft and lids relaxed
Tail loose, relaxed, and upright with a hooked tip
Body leaning into the bunt, with the center of gravity off to one side

ADVANCED CAT WATCHING

Signs of a happy cat

Pet cats relinquish more control than they're naturally comfortable with, and we can unintentionally push their tolerance to its limits. We're all guardians of our cats' happiness, so the more of these heart-warming and satisfying moments we see, the better humans we're being.

Healthy mind

Forward-facing ears and bright eyes with small pupils indicate a positive mood (see pages 12–13). A laid-back posture—stretched out, with paws and tummy skyward—are good signs, too. Greeting you with an upright tail and little purrs, chirrups, and meows also suggests cheeriness (see pages 52–53).

Healthy body

A glossy well-groomed coat, sharp nails, bright eyes, and a clean bottom are all healthy signs. A stable weight, svelte body, and a hearty appetite mean there's a nice balance between in-goings, out-goings, and activity levels. Health, fitness, and freedom from pain and illness (see pages 146–147 and 164–165) are all checks in the happiness box. When vet care is needed, a trauma-free experience will foster happier vibes (see pages 152–153).

Feeling the love

All cats need the opportunity to give and receive affection on their own terms. Many are devoted lap-warmers or chest-kneaders, others are shin-rubbers or bunters, and some leap into our arms purring. Many cats have feline or even canine buddies, too—it's all an expression of love and happiness.

Wildcat pursuits

Cats are happiest when they're marching to the beat of their inner wildcat's drum. Exploring, problem-solving, climbing, and jumping can all provide stimulation and satisfaction. Playing with a runaway leaf, toy, or the real deal (a mouse!) gives a positive rush, while fresh air and sunshine are always welcome—in a garden or catio, or on a windowsill.

Personal space

Desiring home comforts and time and space to do your own thing isn't laziness—R&R is important recharge time. A sunbathing cat with eyes half-open and belly skyward, or one settling down into a postprandial auto-clean cycle, or curling up into such a tight ball that you can't tell which end is which, all epitomize feline contentment.

My cat is a laptop lounger

Whenever I'm reading or working from home, my cat wants quality time with me. They've deleted paragraphs with their paws and sent emails before I'm ready—they literally get between me and my work.

What's my cat thinking?

Jumping up to eye level to be close to you, obstructing your view, and interrupting your activity is attention-seeking behavior. Your cat isn't being demanding, they're just politely reminding you that they need some interaction with you, whether that's a rub under the chin, snuggle time, or a hardcore play session. This can sometimes escalate into vocal demands (meowing), pawing, or knocking things off your desk; if they're really frustrated, they might resort to biting. This doesn't mean they're a "bad" cat or you're a "bad" human, it might just be that you've been too engrossed in what you're doing to have considered socializing with them at the moment.

the function?

Innate curiosity at what's keeping your attention away from them is probably what drives your cat to investigate, as well as the simple desire just to be near you.

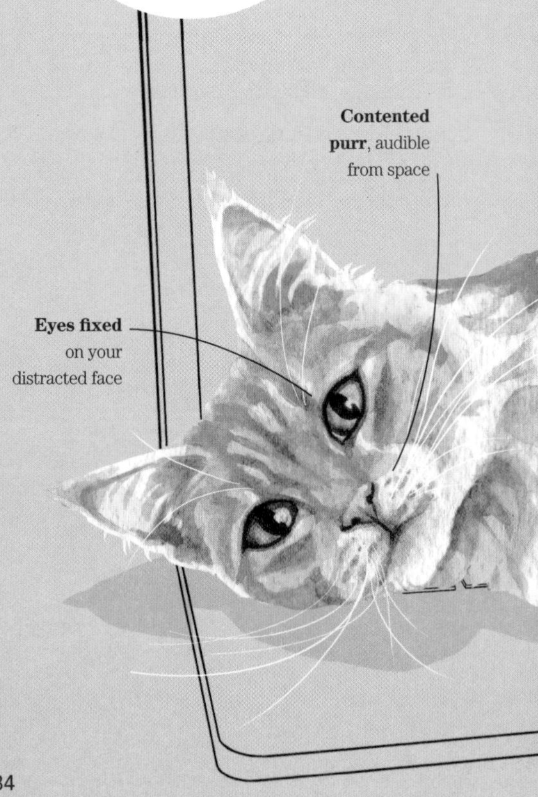

Why cats love tech

There's evidence that animal behavior is affected by electromagnetic fields that emanate from devices, while the heat output is a draw for cats seeking warmth from their surroundings. Also, the screen may offer an array of alluring noises, lights, and a moving cursor.

What should I do?

In the moment:

- **Harsh as it may sound**, ignore your cat's charms in this moment and context. If you plan on getting any work done, don't succumb to their pity-kitty stare or persistent pawing for attention.
- **Check in with their needs**—could they be lacking stimulation and exercise? Occupy them with a toy or a puzzle (see pages 138–139).

In the longer term:

- **Aim for a compromise**—let your cat stay in the room with you but don't allow them on you or the desk.
- **Set up a sanctuary** beside you, with knockout appeal so your cat is happy to settle there. A raised shelf, windowsill, or cat tower—made cozy with one of their blankets or your old sweater or shirt—ensures they can see what you're up to. Mimic the laptop's screen and warmth with a window and a heat pad—few cats can resist either. Seal the deal with a few treats and a water station.

My cat bats their eyes at me

I'm sure my cat is trying to communicate with me by blinking, sometimes with one eye but most of the time with both. I've heard them called "cat kisses"—does it mean they love me?

What's my cat thinking?

If a cat looks at you with relaxed eyes and gives a series of slow-motion half-blinks, then keeps their eyes narrowed or shuts them, you have probably just experienced kitty approval. These "cat kisses" have been likened to the feline version of a human smile. Cats seem to understand our emotional cues and have perhaps worked out that when our eyes narrow into half-moons, as they do with a genuine smile, it's a sign that we're in a good mood and they're matching our vibe. Or maybe, when they trust us, they relax enough to give their usually hypervigilant eyes a break from all the staring (see pages 78–79).

Cat-on-cat blinks

We're not the only ones on the receiving end—cats blink at each other, too, but this behavior often has a very different purpose. When they meet a rival cat, slow blinking indicates their vulnerability and shows they're not keen on all-out physical warfare, whereas a direct stare is used to intimidate adversaries during a confrontation.

What should I do?

In the moment:

- **Repay the compliment**—slow blinking works both ways, so be sure to engage with your cat when they display this behavior.
- **Relax your face** and don't stare directly into their eyes, as this could seem threatening.
- **Check for pain or infection**—half-closed eyes with no blinking can indicate pain (see pages 146–147). If one eye remains shut, or blinking is accompanied by squinting, redness, watering, sticky discharge, or cold or flu symptoms, see a vet. Eye problems can escalate quickly if left untreated.

In the longer term:

- **Connect with your cat** on a deeper level more often by adding slow blinking (with indirect eye contact) to your daily interactions.
- **Make eye contact** on your cat's terms and not when they're showing fear (faster blink rates), frustration, or anger (see pages 102–103), or when they're in play or predator mode.

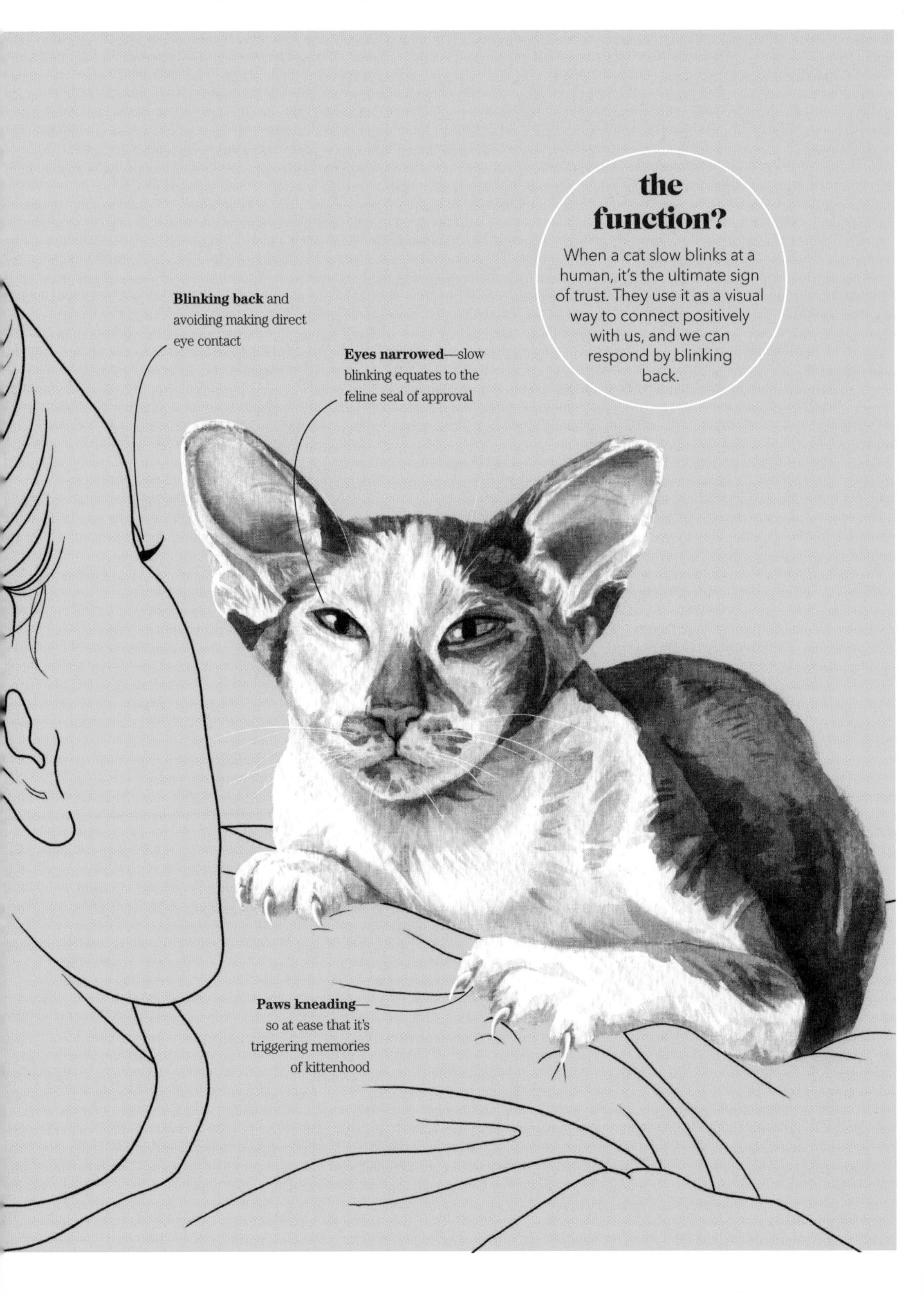
the function?
When a cat slow blinks at a human, it's the ultimate sign of trust. They use it as a visual way to connect positively with us, and we can respond by blinking back.
Blinking back and avoiding making direct eye contact
Eyes narrowed—slow blinking equates to the feline seal of approval
Paws kneading—so at ease that it's triggering memories of kittenhood

My cat hates my new partner

I'm worried my cat has smelled a love rat. Ever since my new partner arrived on the scene, my cat avoids us or hisses. Should I get rid of them? (The partner, not the cat!)

the function?

As territorial prey animals, cats are wary of unknown creatures and identify friends by their familiar smell. A new partner challenges these instincts on many levels.

What's my cat thinking?

Your cat's rejection is not necessarily because they can sense a shady character. Any new person in their territory is a big deal. Cats are sensitive souls who hate change, and new people in the home inevitably alter the routine. You may go out or stay in more, or the newbie may spend the night, along with their unfamiliar-smelling belongings. To earn the feline stamp of approval, your partner must want to win your cat over—and be willing to put in the time. This should help you weed out any unfit suitors.

Avoid using labels such as "naughty" or "spiteful"–your cat is likely feeling confused and anxious, not jealous or plotting your partner's swift demise.

What should I do?

In the moment:

- **Don't force an interaction** or let your partner grab your cat. Ideally, they should avoid physical and eye contact, as this will seem threatening and will confirm they are to be feared.
- **Don't try too hard**—suggest your partner sit quietly on the opposite side of the room, gently toss treats to the cat, and avoid making any sudden noises or movements.
- **Give your cat time**—earning trust takes patience and persistence.
- **Scent swap** (see pages 14–15)— let your cat sniff and investigate your partner and their things, because there'll be lots of new smells for them to get used to.

In the longer term:

- **Help your cat** overcome their anxieties by ensuring your partner becomes irresistible to them. Train them to be your kitty's personal chef, treat dispenser, and game-show host. If your partner is unwilling, it could be time to "swipe left."

Let's talk about sex

Hormonal changes, which cats detect in our unique scent, may affect their response to us—for instance, some cats react positively to pregnant and nursing women. Cats with no experience, or only bad experiences, of one gender may be more anxious around them (see pages 18–19). Men also tend to have deeper, louder voices and bigger feet than women, which may scare a timid cat.

SURVIVAL GUIDE

Vacation cat care

When you're away from home, you want peace of mind that your cat is safe, well, and in good hands. You also want to know that they'll be relaxed and happy with the person taking care of them.

1

There's no place like home

The best place for your cat to be while you're away is tucked up at home, with all their creature comforts. Change, cages, cars, and unknown cats in a cattery all equal stress and potential exposure to viruses. Cats are safer and happier in their own territory.

2

Cat-sitter criteria

Seek out a pet-sitter who's a natural with cats. Check that they're certified (see page 190), insured, and have training in first aid for cats. Good cat-sitters will want to meet you and your cat beforehand–check out reviews, but let your cat decide.

3

Be prepared

Stock up on all the usual food, treats, cat litter, and medications. Ensure you and your vet can be contacted in an emergency. Brief the cat-sitter on any quirks your cat may have–dietary intolerances, habits, and favorite hiding places.

4

Keep it predictable

Set the heating and lighting to go on and off at the usual times. Ask your cat-sitter to pop in at least twice a day and, as far as possible, to stick to your usual timetable for meals, play, petting, grooming, and litter-box cleaning.

5

Home tech

Devices such as timed feeding bowls, automated cat toys, and cameras that link to your smartphone can all provide reassurance for you while you're away from home. However, they should never be an alternative to daily human care.

My cat gives me mixed messages

Like Jekyll and Hyde, my cat rolls onto their back for a belly rub, then snaps into scary mode, grabbing me in a teeth-and-claws death grip.

What's my cat thinking?

It's easy to misinterpret a seemingly relaxed cat, amorously rolling over and exposing their belly, as inviting you to pet them. But this behavior should come with a warning: "Danger, proceed at your own risk!" It's just a cat's way of saying, "I'm pretty chilled out, and I trust you enough to expose my most vulnerable parts"—in no way are they giving you permission to touch. Some cats may tolerate being petted in this situation, a few even appear to relish a tummy rub,

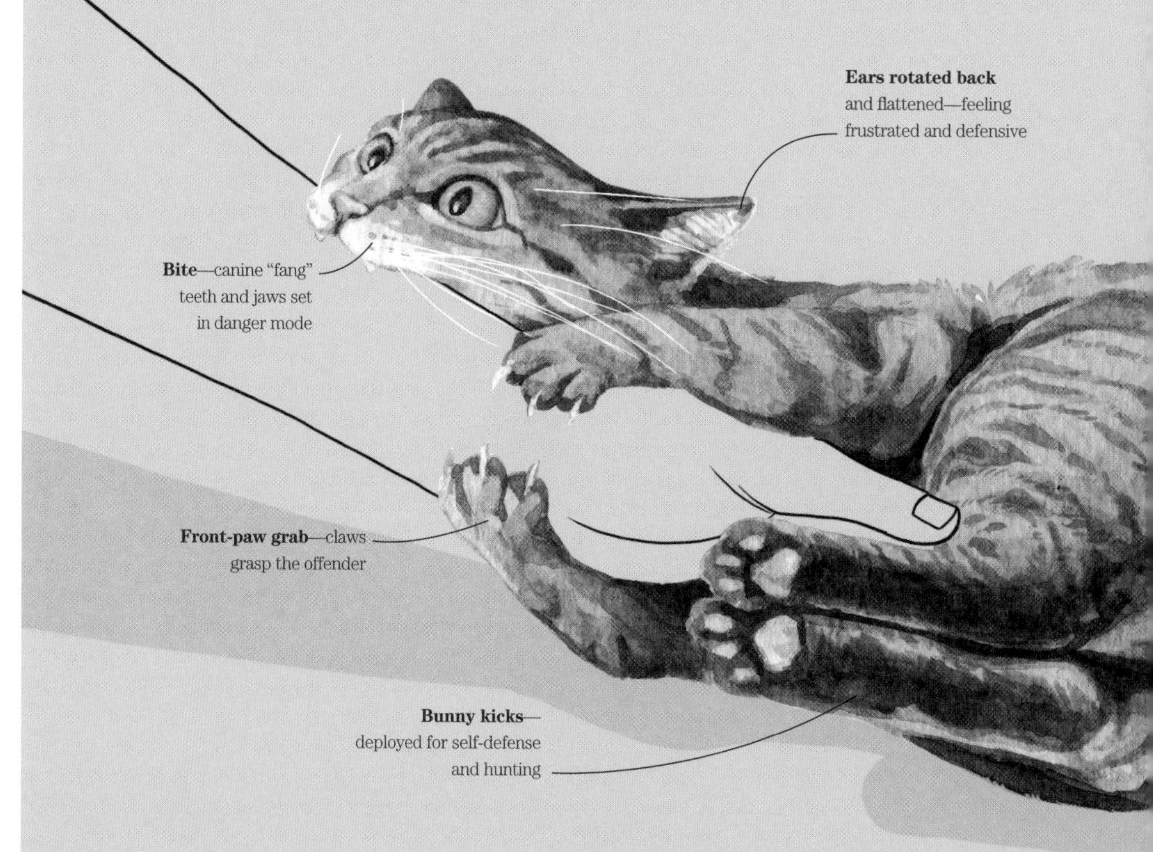

but many will hurtle into defense mode quicker than you can get your hand the heck out of there. The key, when respectfully interacting with cats, is to always let the cat initiate physical contact. Don't say you haven't been warned!

What should I do?

In the moment:

- **Freeze and stay quiet** (see pages 94–95). Move away as soon as possible and don't engage with them in any way while they're so wound up.
- **Don't assume** this behavior is play-motivated. Resist the urge to turn it into a game, which would signal you're cool with it—playtime aggression is fine when it's directed at toys, but not at humans.

In the longer term:

- **Don't do it!** Never push your luck and provoke this defensive armed response.
- **Trust must be earned**, so tune in to your cat's signals and respect their wishes. You'll need to play the long game if you want to see a belly roll.
- **There's a time** and a place for petting. Switch your focus to more appropriate situations and stick to "green light" body zones that your cat is happy to share (see pages 52–53).

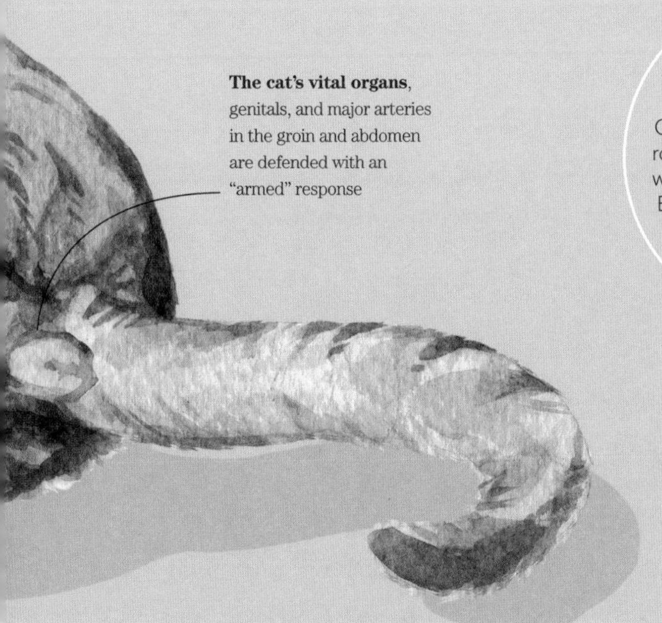

The cat's vital organs, genitals, and major arteries in the groin and abdomen are defended with an "armed" response

the function?

Cats may deploy a full body roll and rub as a greeting, or when scenting or stretching. Exposing their belly signals they're not up for a fight and is a sign of trust.

My cat swipes at my ankles

My "stair beast" sits on a step waiting, and then swipes at my ankles as I pass. Sometimes they even chase me upstairs.

What's my cat thinking?

Your cat is motivated by a normal instinct to stalk and pounce on moving prey. The issue in this case is that you are the unsuspecting target. A cat's eyes are biologically programmed to pick up movement, which is what makes them such successful hunters. It's also why they love pouncing on feet under a duvet or passing on the stairs.

Often, kittens learn from humans that this type of "prey" is fair game—it's cute when they're young, but not so much once they've grown big teeth and killer claws. If your cat gains any reward or adrenaline rush from your reaction, it perpetuates the behavior and risks it becoming an attention-seeking ploy. Add in frustration from a day with little stimulation and you've got the perfect storm—and a scary stair beast on your hands, or ankles.

What should I do?

In the moment:

- **Keep calm and stay still**—if you squeal, try to wriggle free, or run away, you will simulate real prey. This is great fun for a cat and is likely to trigger a pursuit or a bite.
- **Remove yourself** from the situation (and your cat) as soon as possible.
- **If your skin is broken**, rinse well, and seek medical help for any deep scratches or bites.
- **Note the timing and context** of your cat's response. Have you come home covered in new smells? Has there been tension with other cats?

In the longer term:

- **Schedule play sessions** to help your cat stay in tune with their inner wildcat and let off steam safely, using fishing rod and laser toys that distance you from the "prey."
- **Provide solo fun** with puzzles and self-play toys (see pages 46–47, 138–139, and 182–183); if possible, allow access outdoors (see pages 64–65).
- **Don't mislabel your cat** as aggressive—they are likely acting out a normal behavior that has been misdirected at you, which is a potential red flag that all is not as it should be in their world.
- **If you're worried**, call the vet—if it's more than predatory play, it may signal pain or illness, so assessment and referral may be needed.

> Yelling or shooing are not helpful responses and will only add fear to an already complex array of emotions.

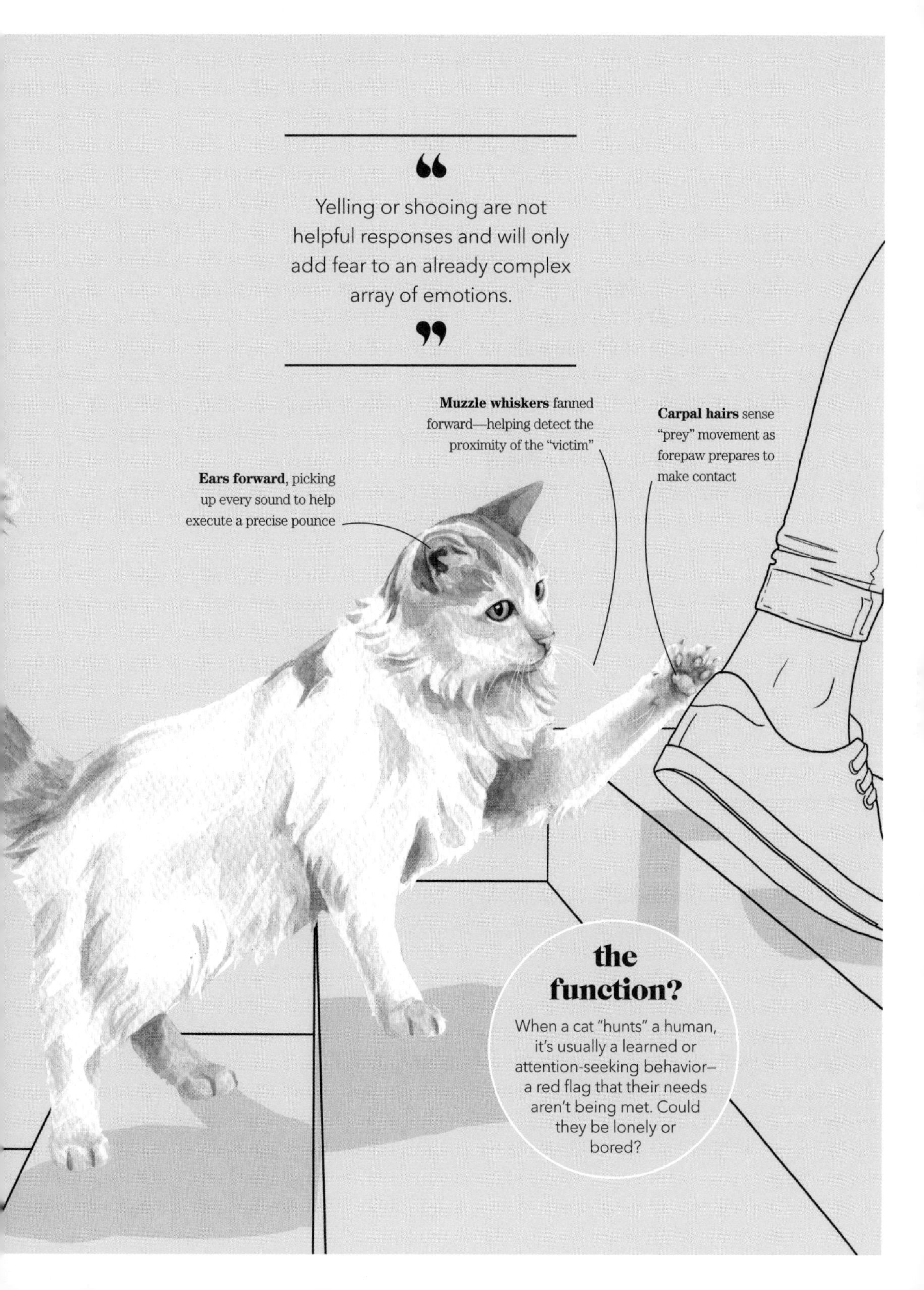

the function?

When a cat "hunts" a human, it's usually a learned or attention-seeking behavior–a red flag that their needs aren't being met. Could they be lonely or bored?

SURVIVAL GUIDE

Cats around kids

Always supervise interactions between your cat and the next generation of feline fans, so you can guide kids on how to safely gain a cat's trust and how to spot if they're feeling uncomfortable.

1

Choose wisely

Timid or shy cats tend not to enjoy life in busy, noisy family homes, whereas calm, social, playful, and less fearful cats fare better. Before you adopt a kitten or cat, ask about their past experiences with kids.

2

Monitor and guide

It's essential to keep a close eye on the child's behavior and the cat's body language. If a cat lashes out, it's because you've missed all the early signs of their uneasiness (see pages 102–103 and 122–123).

3

Show respect

Teach children how to respect a cat's choices and safe spaces, and to check with you before interacting. Encourage gentle strokes with the back of the hand, and don't let them grab, poke, pick up, or chase.

4

Offer cats kid-free zones

Cats don't appreciate noise, unexpected movements of toys, or the unsupervised attentions of young children. Create retreats high up, so they can escape the busy floor level, and provide quiet, safe places (see pages 46–47) where they won't be disturbed when resting or using litter boxes or food and water stations.

5

Create positive memories

Children love games and feeding animals, and cats like to play and eat, so embrace their compatibilities with fishing-rod toys, bubbles, and treats (see pages 182–183). Equally, children's bedtime stories, movies, and snuggle times can be opportunities for cats and kids to sit calmly in the same room.

My cat tails me to the bathroom

I love that my cat follows me everywhere—and I mean everywhere. If I nip to the bathroom, they come, too, and jump onto my lap; sometimes they even sit in my pants! Too much information?

What's my cat thinking?

Despite having a reputation for being detached and uninterested in their humans, cats are actually very curious about what we do. Bathrooms are full of interesting things such as running water, rolls of paper, cool ceramics—and, when you drop your trousers, warm skin and fabric that smells of you. Your cat probably just wants to mix scents and reaffirm your bond.

The bathroom can be a refuge from the chaos of kids, other pets, or grumpy adults. Maybe your cat shares your need to shut the hectic world out and wants a slice of the inaction, too. Some cats just hate being on the other side of a closed door (see pages 118–119).

The bathroom may not hold the same appeal for you, but it's cool, calm, and full of fascination for a feline–especially when you're there, too.

What should I do?

In the moment:

- **Make a choice**—are you happy to let your cat in the bathroom with you or would you rather keep them out?
- **If you're not shy**, make sure it's safe—with no hot bath for them to fall into.
- **If you prefer privacy**, you may need to put up with some protesting.

In the longer term:

Get ready to multitask! Plan ahead so you can put your private time together to good use. When you're anticipating a seated performance, forgo the book and try some of the following:

- **Schedule some playtime** around your bathroom routines—keep a fishing-rod or laser toy on hand.
- **Cats are smart** and learn from us every day (see pages 132–133). Tap into this and teach your cat to sit, raise a paw, or play fetch.
- **Have a kitty spa session**—use the time to groom that silky coat or massage their worries away.
- **Make a fuss** over your cat and pet them in all their favorite spots.

Tail-up greeting—
"Need a lap warmer?"
the
function?
This is likely to be the feline equivalent of a dinner à deux: quality time with you. Unless it's just curiosity about that oversize teacup where you mark your territory.
Expectant gaze—willing you to put your book down
Purring loudly—in anticipation of lap time

My cat's a petty thief

My cat can't resist stealing socks, soft toys, rubber bands, even the dish sponge. They carry them in their mouth, wailing. Are they a feline retriever or do they think these things are kittens?

What's my cat thinking?

It may seem as though your kitty has lost the plot when they start fetching random items from around the home. They're unlikely to be "thank you gifts" for being such a great human. A more plausible explanation is that this is your cat's quirky way of engaging with their natural instinct to hunt, even when there's no real prey to be found. Sometimes the catch of the day will be a toy, which they may start hiding in a favorite spot. Some female cats shift their stash from time to time, as they would a litter of kittens.

Perhaps we mislabel this behavior as theft because the items seem to be taken covertly, are missing in action, or reappear in odd places. But this makes sense, as most hunting occurs between dusk and dawn, when we are likely to be distracted or asleep.

Why the vocal performance?

The distinctive wailing meow ("oooh-wow") communicates a catch. For millennia, a cat's worth to humans depended on their hunting prowess. Vocally proclaiming a rodent kill probably ensured praise or a tasty tidbit as a reward.

What should I do?

In the moment:

- **If you're OK** with your cat displaying their natural urges, make a fuss to communicate your approval.
- **If you want to discourage** the activity, just ignore it—even if they haul your lingerie into public view. Never reprimand or punish a cat for the "crime" of being a cat.

In the longer term:

- **Play to their strengths** and tap into their inner predator through play, using toys they can hunt and capture—ideally ones that include their preferred texture or smell.
- **Redirect the behavior** with alternative items—for instance, you could donate a pre-loved pair of your socks and make them more exciting by putting catnip in the toes.
- **Reduce opportunities** to "steal" by picking up your dirty laundry!

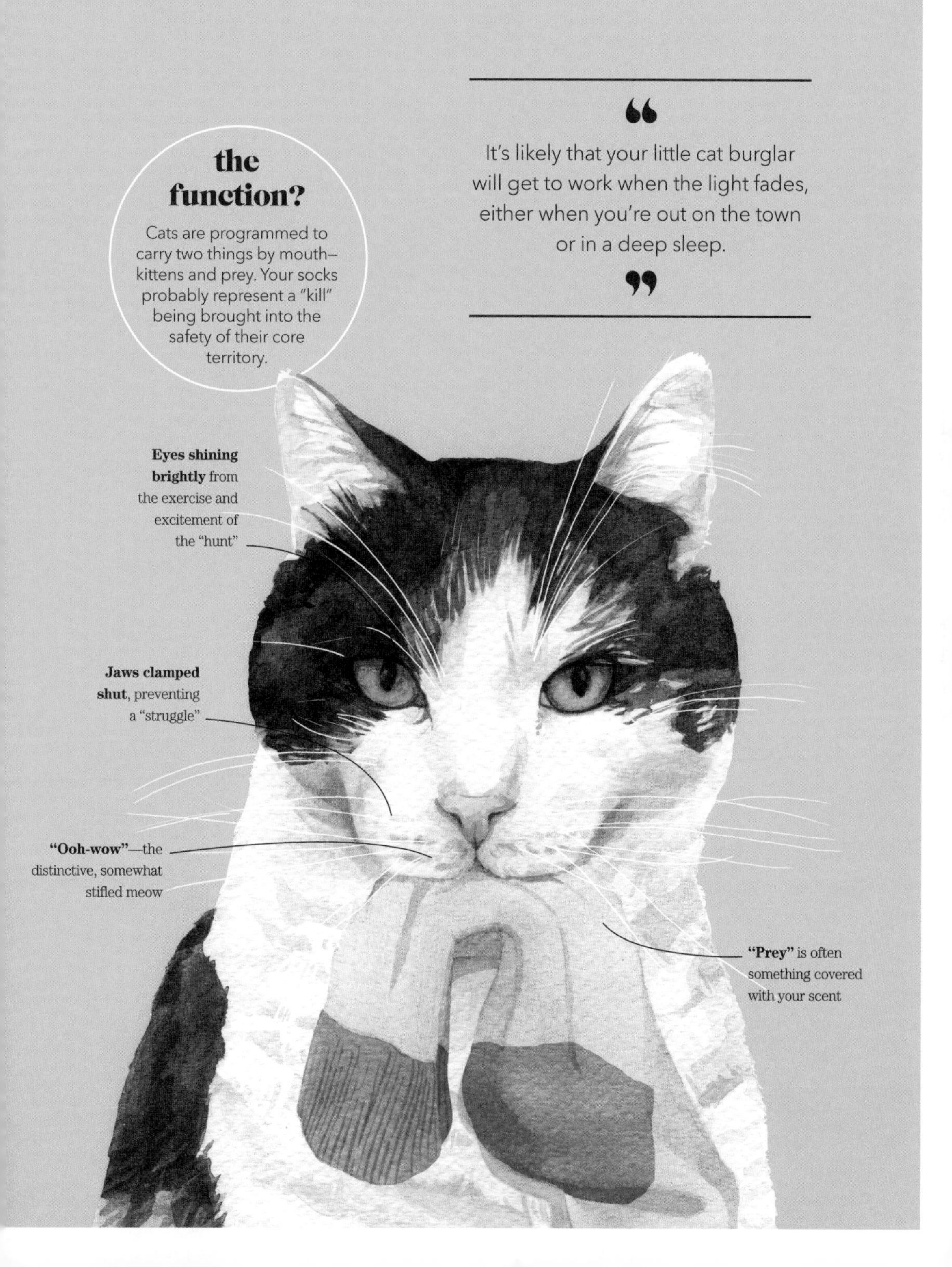
the function?
Cats are programmed to carry two things by mouth–kittens and prey. Your socks probably represent a "kill" being brought into the safety of their core territory.
It's likely that your little cat burglar will get to work when the light fades, either when you're out on the town or in a deep sleep.
Eyes shining brightly from the exercise and excitement of the "hunt"
Jaws clamped shut, preventing a "struggle"
"Ooh-wow"—the distinctive, somewhat stifled meow
"Prey" is often something covered with your scent

ADVANCED CAT WATCHING

Signs of a grumpy cat

There are no aggressive or psycho cats—or even grumpy cats—just cats feeling threatened. We all act on raw instinct when gripped by fear, and lashing out is a last resort—a survival tactic to cope amid a whirlwind of negative emotions.

Don't unleash grumpy cat

Vets know better than most that trying to get hands-on with a cat against their will during a time of fear, pain, or illness can unleash a volatile mix of emotions (see pages 30–31). If you ignore the signs that your cat is scared (see pages 122–123), it won't be long before fear meets frustration, frustration turns to rage, and grumpy cat emerges. Before you know it, you'll be heading to the ER. Always get a vet to check a grumpy cat for pain or illness, and to give well-being advice.

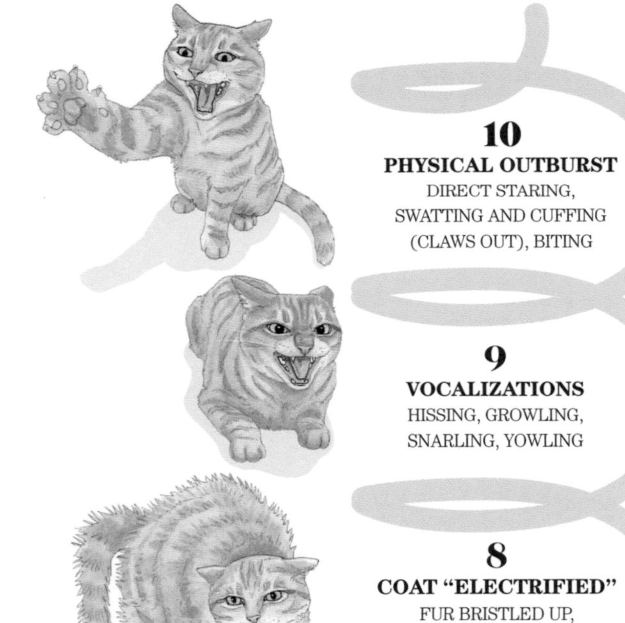

The coiled spring

It's easiest to understand how things can spiral out of control using my analogy of a coiled spring. A stressed cat is like a tightly coiled spring under tension. When pushed too far, they can explode in an instant, releasing all their potential energy—at you! Each coil of the spring represents one of the many feline visual or vocal protests that you've missed. Not all cats show all the signs all the time, or in a strict order, and they're lightning fast, so now is not the time to blink.

Learning the hard way

Some cats have learned through bad experiences with humans that violence is the answer. Or, at least, it gets people to stop what they're doing and think twice about doing it again. But it saves a whole lot of stress if you think before you start.

Compressing the spring

Each coil of the spring shows clues (see pages 12–13) that a fearful cat (1–4) has also become agitated and frustrated (5–9). This emotional mix ultimately unleashes grumpy cat (10).

My cat isn't a people pleaser

My dog bounds to the door to greet me the instant I get home, but my cat barely opens an eye—until I rattle the food.

What's my cat thinking?

While some cats are very affectionate, others are less demonstrative. Your cat probably is pleased you're home—just not enough to cut short their nap and vacate their cozy, warm spot to come and greet you. It's easy to forget that felines haven't been bred for millennia to work or keep us company in the way that dogs have. Cats have only been living indoors with us for around 150 years, so they aren't as proficient at reading our body language, or as interested in our every move. They don't know we need their esteem—they certainly don't need ours!

the function?

Pet cats may depend on us for life's essentials but they're still free-spirited. Their inner drive to please themselves isn't arrogance, it's a survival instinct.

Going it alone

Cats are programmed to hunt alone—caring for other creatures, unless they're offspring or close relatives, brings no benefit. Pet cats override their wildcat heritage to varying degrees. Some will socialize and bond with us, other cats, or even a D. O. G., but this very much depends on their genetics, breed, temperament, and life experience.

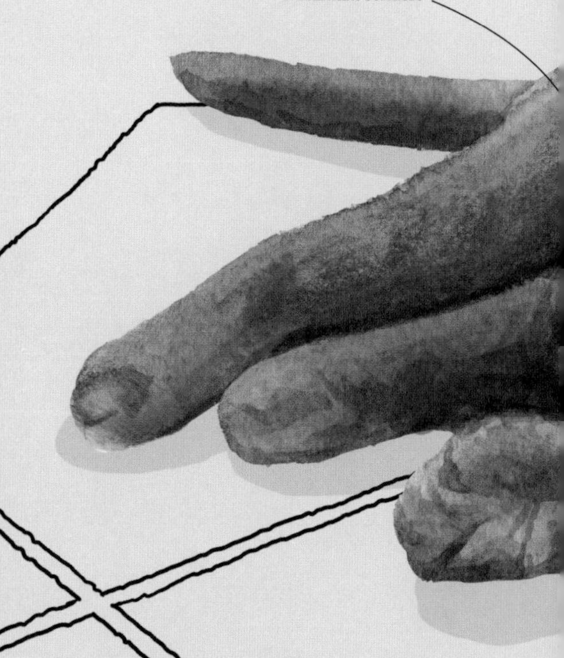

Sprawled out for maximum comfort

What should I do?

In the moment:

- **Respect your cat's needs** and wants. Interact on their terms and let them enjoy a well-earned rest.
- **If the behavior is new**, get a vet check. Lethargy, depression, and avoiding interaction or hiding can all be signs your cat is anxious, in pain, or feeling unwell (see pages 122–123, 146–147, and 164–165).

In the longer term:

- **While some cats greet** their humans with the enthusiasm of the extroverted, tail-wagging dog, the affectionate gestures of many cats are much more subtle. Look instead for cat love clues, such as trills, meows, purring, bunting, rubbing, or choosing to be in a room with you.
- **Whenever you come home**, encourage your cat to come to you for treats and play (see pages 132–133); they may start to see your homecoming in a new light.
- **Schedule some** quality one-on-one time with your cat.

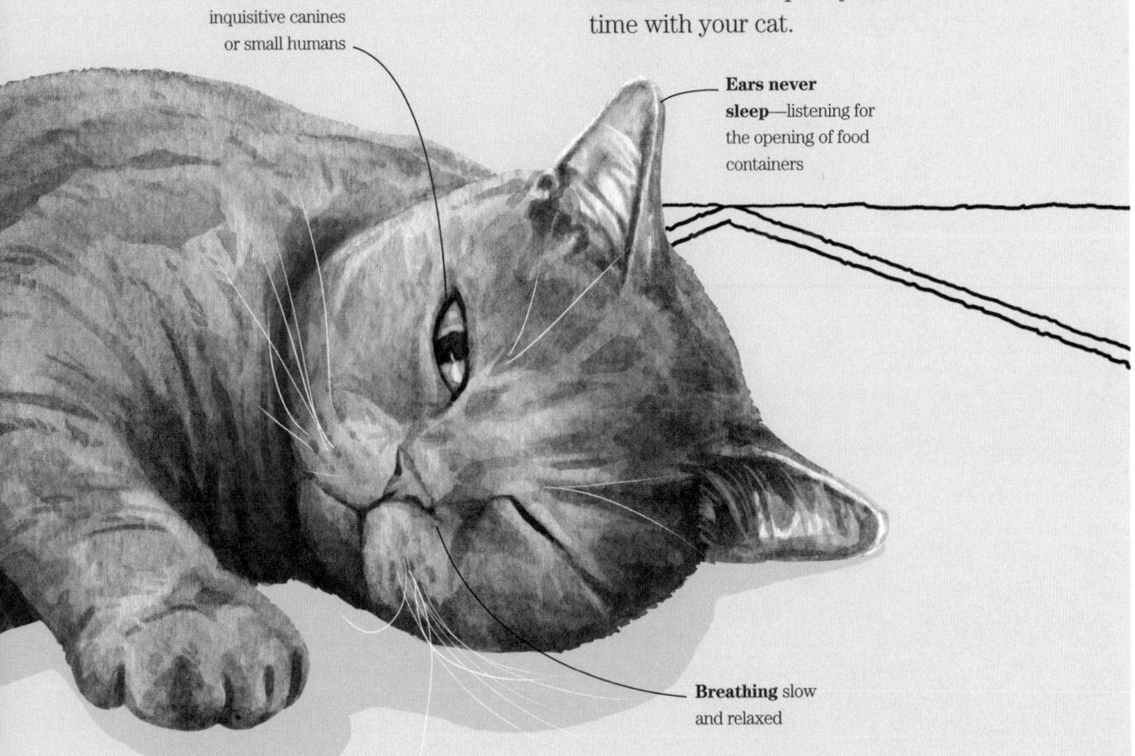

CHAPTER THREE

My cat drives me crazy!

We love our cats, but sometimes they do things that exasperate us. Keeping an open mind and looking beyond your cat's annoying habit to the instincts and emotional motivations behind it will help you resolve the issue and improve their well-being.

My cat wants to play at 4 am

My cat is laid-back all day, but in the early hours they suddenly decide it's time for play—and breakfast. How can I get them to let me sleep in peace?

What's my cat thinking?

Your cat comes from a long line of desert-dwelling ancestors who were most active between dusk and dawn, when the temperature was cooler and their rodent prey was most likely to be about. Although living with humans has shifted cats' activity patterns more in line with ours, this is still the natural time for a cat to be most active. And, since they are full of energy, it seems only logical to them that they should let you in on the fun, too. If your cat is bored and frustrated, they may look to you to entertain them. And perhaps there have been times you've given them treats to distract them while you staggered back to bed. If so, your cat will assume, understandably, that a 4 am snack is always a possibility.

What should I do?

In the moment:

- **Don't interact at all**—any communication, even if it's to shoo them away, is rewarding them with a reaction. Your cat needs to know that their behavior will not reap rewards. It may not be easy, but it will allow you to regain your vital shut-eye, so stand your ground.

In the longer term:

- **Schedule extra playtime** an hour or two before you go to bed. Make it a full-on hunting, stalking

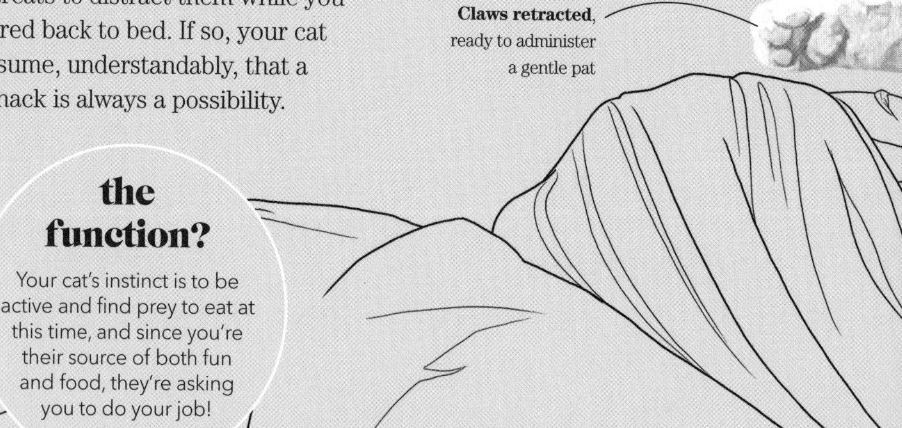

Claws retracted, ready to administer a gentle pat

the function?

Your cat's instinct is to be active and find prey to eat at this time, and since you're their source of both fun and food, they're asking you to do your job!

experience (see pages 70–71) to give them a fun, physical workout. Aim for 5–10 minutes of vigorous play and repeat if they're still keen. Wind the session down gradually or you risk leaving them wired.

- **Give your cat a meal** after the play session. If you normally feed them twice a day, split the same amount of food over at least five meals instead.
- **Cats thrive** on predictability. The more you keep to regular times for food, grooming, play, and sleep, the more likely they are to feel settled and follow your preferred routine.

Have some special toys that only come out at night–in another room.

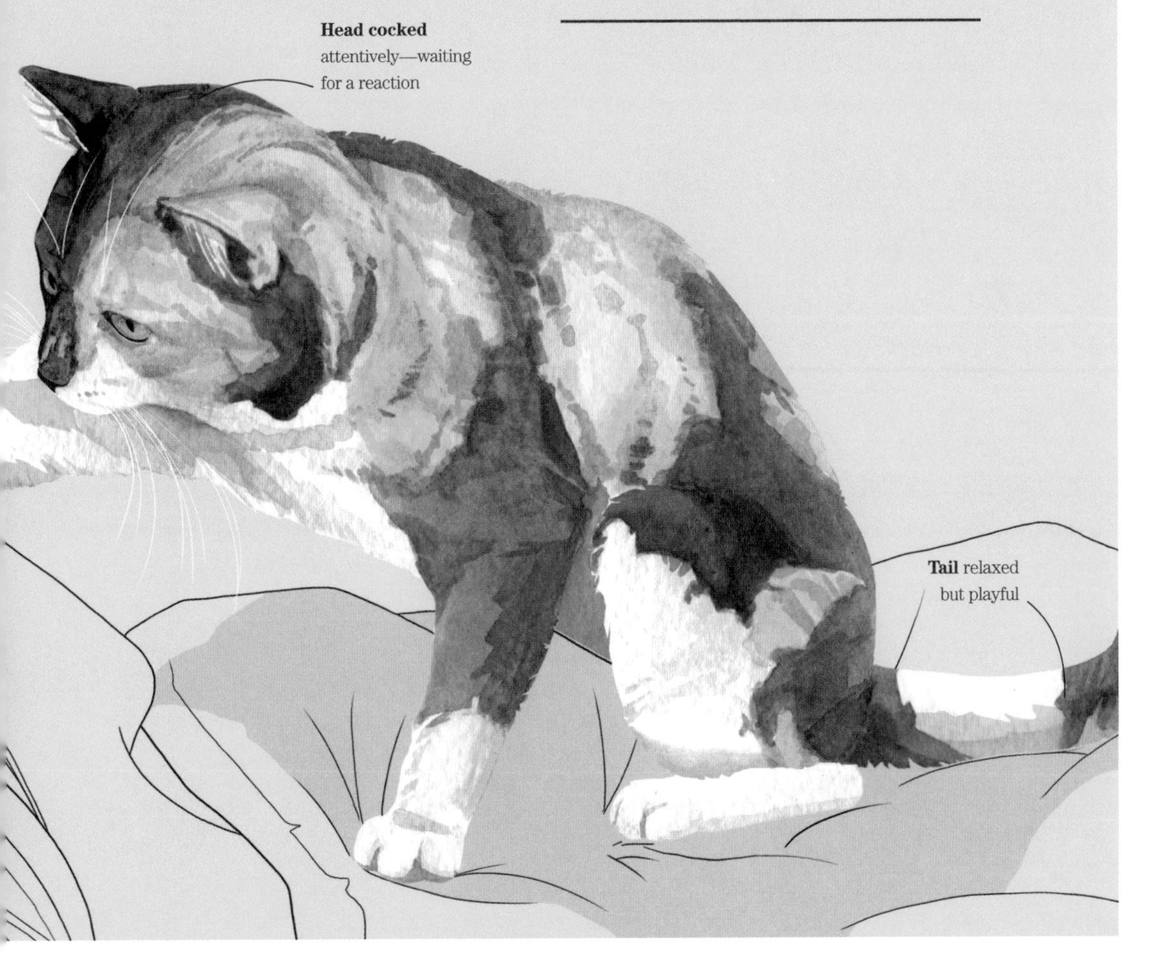

My cat is so greedy

They devour their own food as fast as they can and then muscle in on my other cat's bowl. The slow eater gives in and stalks off in a huff. Is this a case of feline FOMO?

the function?

Cats are solo hunters and opportunistic feeders; they are not natural sharers, and if there's food available, they'll probably eat it. It's all about survival of the fittest.

What's my cat thinking?

The feline digestive system is geared to eating mouse-size meals every few hours. If you only had access to food when a busy human remembered to feed you, you'd be ravenous, too! There are no table manners in a cat's world, so any food that's available is theirs.

Eating is a private affair and makes cats feel vulnerable, so having a fellow feline in direct eyesight is threatening, even if it's only a timid sibling. It also turns feeding time into a competition—whoever gobbles fastest wins, or pukes it back up, but that's another story (see pages 166–167).

What should I do?

In the moment:

- **Make sure every cat** in your household is getting what they need. Under- and over-feeding can both cause problems (see pages 160–161).

In the longer term:

- **Make your cats** feel more secure by keeping them out of sight of each other at mealtimes. This minimizes gobbling and bowl-swapping. If you shut them in separate rooms, set a timer so you remember to open the door 15 minutes later.
- **Microchip-activated bowls** can be programmed to open for specific cats and close again when they walk away, helping ensure cats don't feel rushed and can come back and forth for food as they wish.
- **Predictable feeding routines** that mimic a cat's natural rhythm are reassuring and reduce frustration. Five or more smaller meals a day are ideal, so timer bowls and puzzle feeders can really help with this.

An increased appetite can signal illnesses such as diabetes, overactive thyroid glands, gut disease, and parasites–time to book a vet check.

The hunger games

Some cats get "hangry" when they're overdue a meal and try to assert their right to food over another cat. Staring, blocking another cat's access, and bowl-swapping are the equivalent of pushing to the front of a café line.

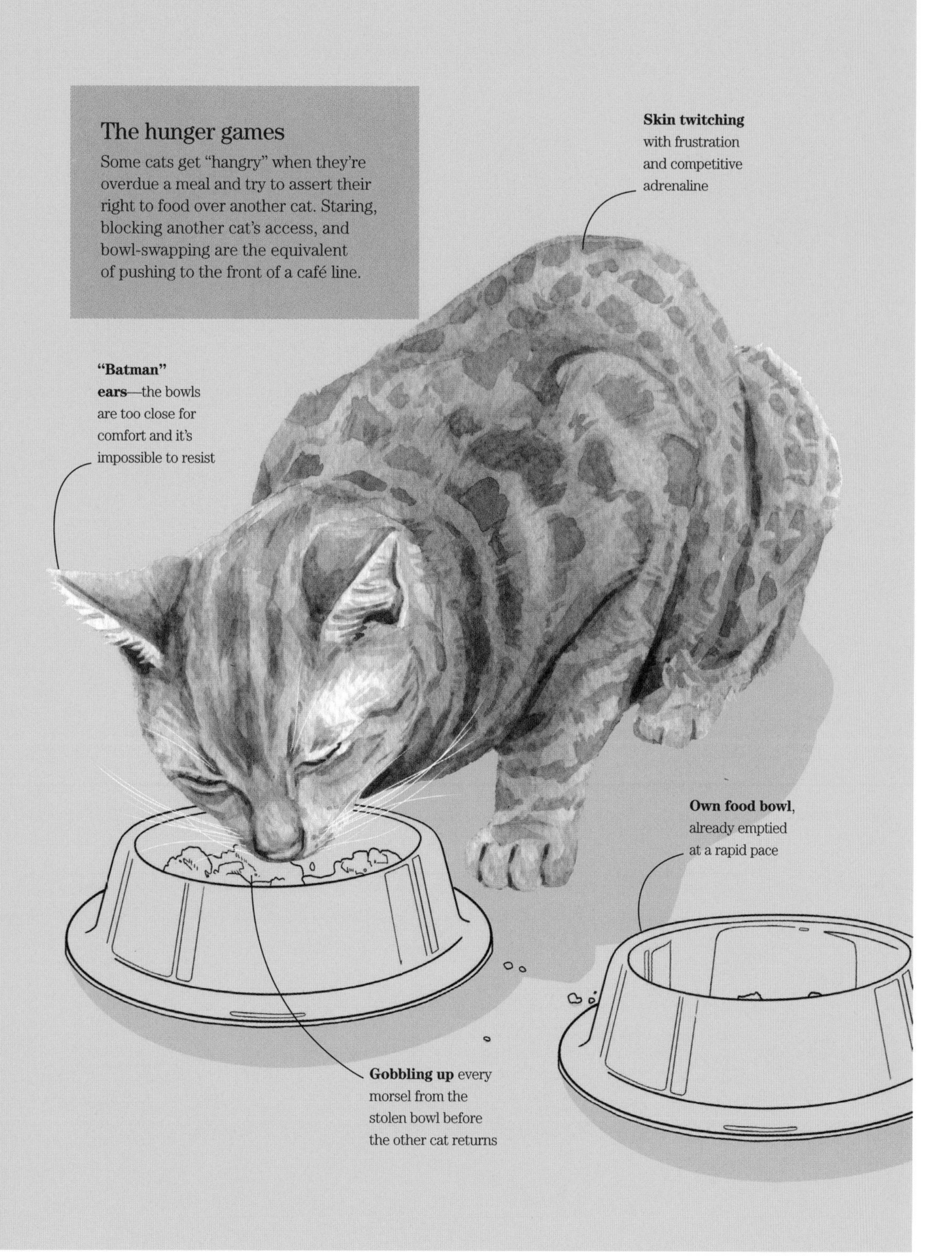

SURVIVAL GUIDE

Introducing new pets

Cats can be best buddies with other cats and even dogs, but first impressions count. You can help smooth the way to a happier long-term relationship by ensuring positive first encounters.

1

Opposites don't attract

Match pets based on their temperament, energy levels, and social skills, not aesthetics or your desires. Think, too, about whether a new pet could represent a potential predatory threat or conflict.

2

Degrees of separation

Optimize the resident cat's habitat and initially separate pets, each with a safe zone (see pages 126–127). Separate them when you're not supervising, or if either pet shows fear or aggression.

3

Take it slowly

First, swap scents (see pages 14–15). Then feed each pet on either side of a closed door. Repeat several times before introducing sight with a mesh screen or pet gate. Supervise until they're relaxed enough to eat without a barrier.

4

Keep your dog under control

Put your dog on a leash during interactions on the opposite sides of a pet gate, and don't let them lunge. Train them to be calm, quiet, and keep their distance. Use toys or treats to distract them and keep their focus on you. Praise and reward calm behavior with more treats.

5

Patience and respect

Never force, restrain, or cage a cat during an introduction. Offer an easy escape route. Cats don't forget bad experiences and they will influence future interactions. Watch their body language closely. Don't rush; let your cat set the pace—this process could well take weeks or months, not days.

My cat bullies the dog

My dog's such a wimp! My cat just gives them "the look" and they get straight out of their cozy bed. The victorious puss then settles down in the warm spot.

What's my cat thinking?

Cats aren't bullies—they're not motivated by getting one-up on anyone, even pesky dogs. They're wired to see dogs as predators and dogs are wired to see cats as prey, so there's bound to be conflict, especially over cozy sleeping spots. As far as your cat is concerned, your entire

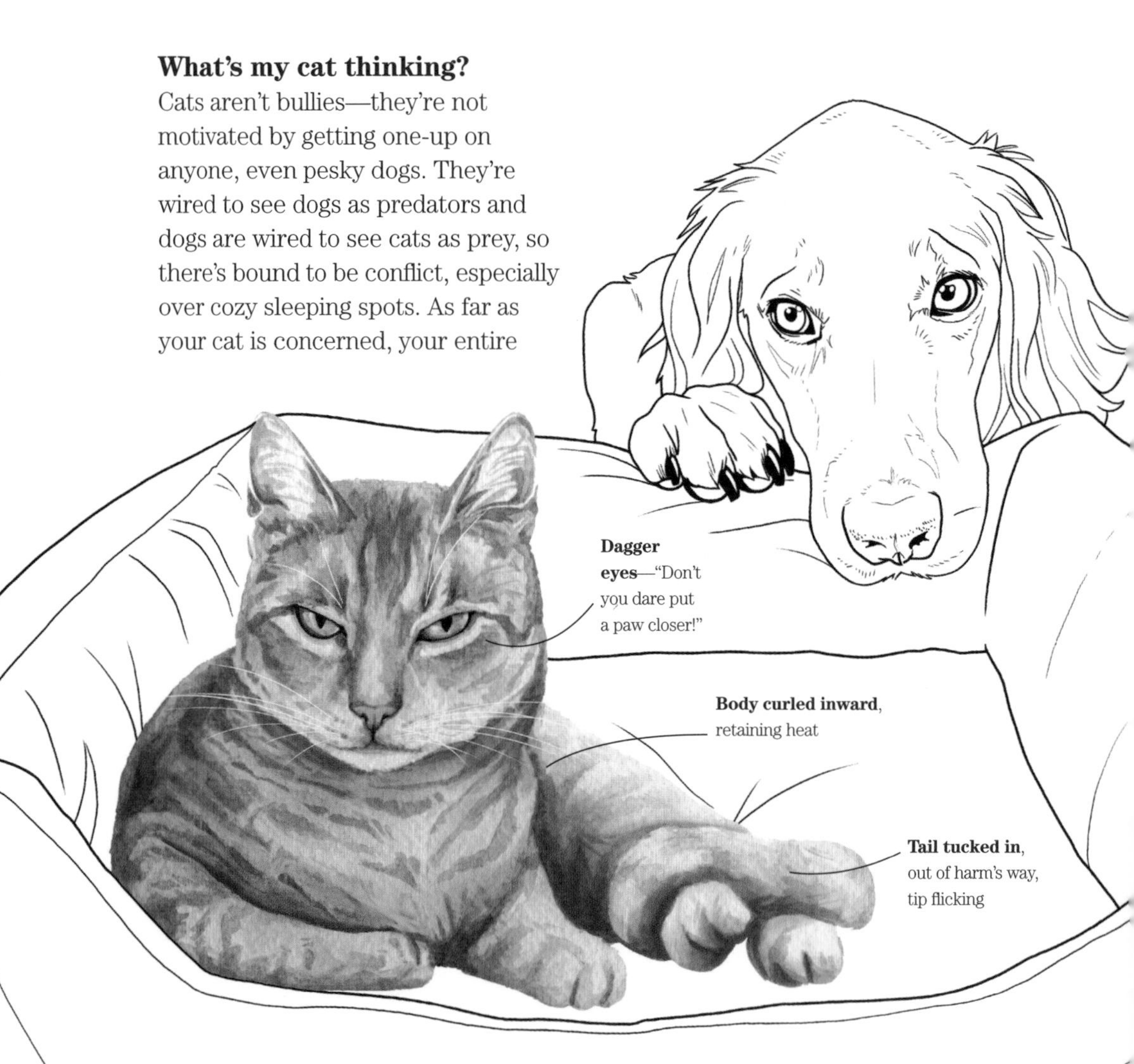

Although dogs are cats' natural predators, a dog and a cat with the right temperaments can live together in a degree of harmony, if introduced in the right way (see pages 112-113).

home is their domain—including the dog's bed. Why wouldn't they want to get their paws on a cozy, pre-warmed spot? Having learned to read your dog's behavioral patterns, your clever kitty has gained confidence and stands their ground, rather than running or freezing, as they may have done in the early days. Now, instead of resorting to hisses or a claws-out cuff on the ear, they can nab that sought-after spot by simply deploying "the stare." Your dog's not as goofy as they appear, remembering that displacing the cat can have sharp consequences.

What should I do?

In the moment:

- **Avoid physically intervening**, as you may tip a carefully balanced situation over the edge. Lure your cat away with a treat or a toy, freeing up the bed for your dog.
- **Assess the appeal** of that particular spot for your cat. Is it next to a radiator or a window? Could you relocate the cat's bed to somewhere just as inviting, and make it even more tempting to them (see pages 54–55)? Or is your cat struggling to access beds and sofas due to pain (see pages 146–147)?

In the longer term:

- **Praise and reward** all positive interactions between cat and dog.
- **Provide plenty** of desirable snooze spots for your cat, including high-up so they have somewhere to escape to.
- **Use cat-friendly pet gates** to reduce tension, prevent "games" of chase, and assure your cat of an easy escape route.

Cat-compatible canines

Many, at least initially, fight like, well, cat and dog—and it's usually the cat that comes off worse. Any dog has the potential to kill, but breeds such as terriers and hounds instinctively chase and bite small furry things that move, so are not a good choice. Others can learn to timeshare a home, and some even embrace each other's company, especially when introduced as puppy and kitten. Breeds such as retrievers and poodles are more likely to be cat-friendly, but a dog's individual temperament and experiences count, too.

My cat sucks and chews odd things

Their favorite pastime is sucking my woolly sweater or fleecy bathrobe, but they've even chewed my laptop cable!

the function?

This behavior could fulfill a number of functions–from alleviating boredom to self-soothing or relieving toothache–so you may need to investigate further.

What's my cat thinking?

This behavior could be an indication of boredom or stress (see pages 122–123 and 162–163). It may also be a sign of pain in teething kittens or in cats with gum disease or tooth decay.

Often it's a throwback from early life. Kittens who are taken from their mother before eight weeks of age retain their suckling instinct and often seem to gain comfort from latching on to things, as they would a teat—just like a thumb-sucking baby.

Some cats can suffer from ritualistic behaviors or obsessive compulsive disorders. Increased hunger and pica (compulsive eating of things other than food) can be seen in cats with nutrient deficiencies or illness.

Chewing and sucking is usually harmless and soothing for cats, but it could cause injury—especially if the target is a live electric cable. Swallowing items such as plastic, paper, or sticky tape could also lead to damage to, or obstructions in, the mouth and/or gut.

What should I do?

In the moment:

- **Let them be**—yelling or thundering across the room will only scare your cat and not deter the behavior. Adding negative emotions can make stress responses worse.
- **If the target** is expensive or hazardous, or your cat is trying to eat it, lure them away with a toy or a treat—but take care not to reward them for the act you want to deter.

In the longer term:

- **See your vet** to exclude medical causes of this behavior.
- **Discourage your cat** by hiding items you don't want them to chew, or by applying safe but vile-tasting bitter-apple spray.

- **Offer a safe alternative** that replicates your cat's preferred target—a soft toy or your old fleecy dressing gown. Make it more enticing with pheromones or catnip.
- **Reduce boredom** with regular play sessions, rotating toys to keep them novel and using mind-stimulating puzzle feeders (see pages 138–139).

All in the genes?

Some breeds, such as Birmans and Siamese, show a genetic tendency for sucking and chewing inanimate objects, especially woolens. The behavior may be triggered by a texture, taste, or smell.

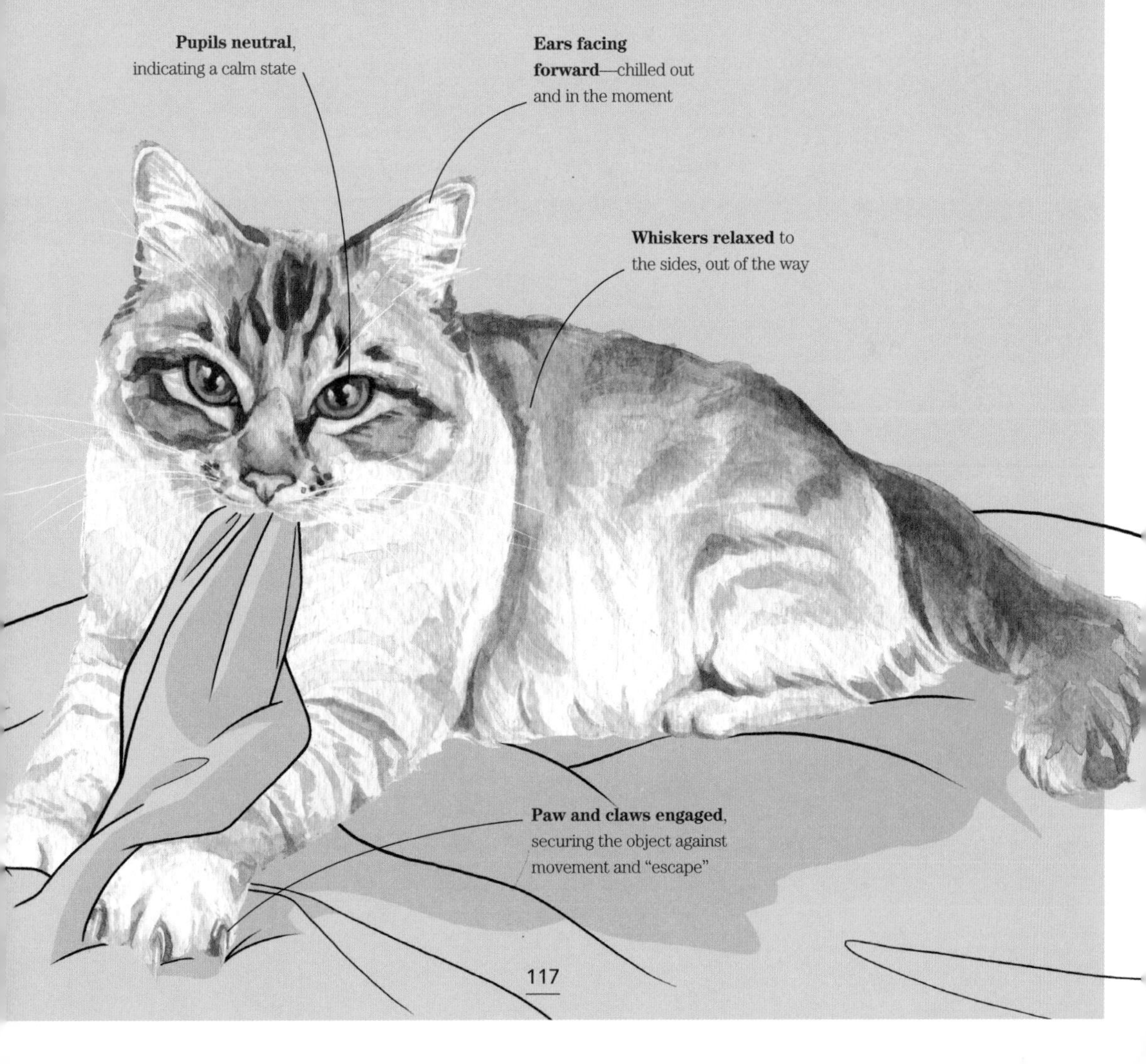

My cat hates closed doors

My cat's a prima donna when it comes to a closed door. They meow incessantly until I open it, then they'll casually poke their head into the room, sniff, and saunter off in the opposite direction.

What's my cat thinking?

Think of the biggest busybody control freak you've ever met and imagine them being forced to live, completely out of their comfort zone, with someone who doesn't speak their language. This is the plight of the domestic cat. It's not hard to see why your cat might be preoccupied with where you are and what you're doing, especially if you close a door.

Cats are naturally curious creatures, especially when there's something new or different. They like to have options, prefer life to be predictable, and want things to be under their control. Your home is also their territory and they have an innate need to micromanage every aspect of what happens there.

Cats aren't mind-readers. You may know the door is going to be opened again soon, but cats don't realize it's only a temporary barrier. As far as they're concerned, it's a permanent access-obstruction that disrupts their routine and undermines their feelings of control and security. It's like waking up to find that someone has built a wall right through your living room.

What should I do?

In the moment:

- **Don't punish or reward** your cat by shooing them away or opening the door, or they will persist in this behavior to get your attention.
- **Think about context** and motive; a single cat may feel abandoned, while the dynamics in a multi-cat household may be challenged if the size of shared territory is reduced.

In the longer term:

- **Think like a cat** and avoid closing doors if you can.
- **When door-closing** is a matter of safety or privacy, plan ahead and distract your cat away from the door, or set up an inviting sanctuary elsewhere (see pages 126–127).
- **Offset feelings of insecurity** by enriching your cat's territory (see pages 46–47).
- **Ensure a closed door** never cuts off access to litter boxes, water, food, and favorite nap spots.
- **To discourage** your cat from damaging the door by scratching, see pages 134–135.

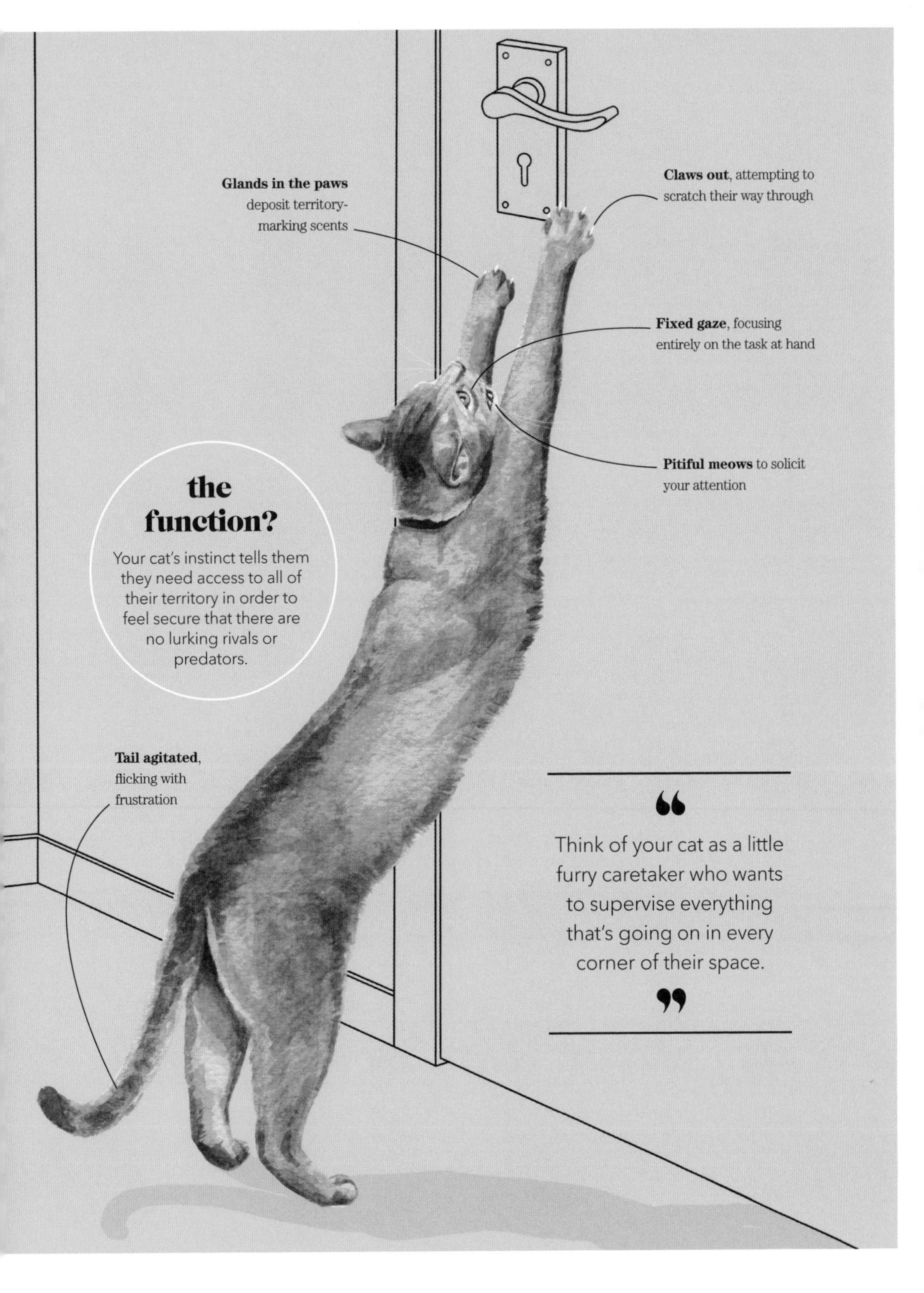

the function?

Your cat's instinct tells them they need access to all of their territory in order to feel secure that there are no lurking rivals or predators.

> "Think of your cat as a little furry caretaker who wants to supervise everything that's going on in every corner of their space."

My cat sprays in the house

My living room stinks of cat pee! Bizarrely, the culprit is a neutered female—I've caught her in the act, spraying the glass doors and new curtains. #MakeItSTOP

the function?

Wildcats spray urine to mark their territory as a social-distancing strategy to avoid physical conflict–the urine lingers so the depositor doesn't have to.

What's my cat thinking?

Cats use urine spraying to mark their territory in their absence and avoid all-out warfare with rival cats. Urine also advertises vitality and sexuality to potential mates. How fresh the spray is indicates when the feline "tag artist" was last there. This is usually the behavior of unneutered males and, occasionally, females.

Context is key. When a neutered cat sprays indoors, it's telling a story of unhappiness, stress, or illness. It tends to happen when multiple cats share a space—more cats mean more stress and higher chances of an aromatic mural. The urine marks result in further stress, so it's a vicious circle.

What should I do?

In the moment:

- **Keep your cool**—if your cat is already insecure, you'll add to their worries if you freak out. It's not fair, polite, or helpful to rub their nose in it, no matter how upset you are.
- **Clean up well and fast**—even a trace of residual pee odor will attract respraying and be threatening to fellow cats, who may add their own "tag" into the mix. UV-light will help show up old, dried urine and reveal the full extent of the damage. Avoid using bleach and ammonia-based cleaners, which smell like cat pee to feline noses, and phenol-based detergents, which are toxic to cats. Products made specifically to clean up urine work best.

In the longer term:

- **Prioritize a vet check**—one in three cats who spray in the home have a medical issue.
- **Identify the cause** of stress or any sources of conflict, such as a change or a lack of control. Are there any aspects of your cat's world that could be improved? (See pages 46–47.)
- **Change the channel** on the "window TV" (see pages 54–55 for some ideas) if strange cats come by.

Feline CSI

Time to turn detective to solve this kitty crime. When a single cat shows signs of stress it's often because their territory is under threat. That neighborhood "trouble-making" cat is the prime suspect, so review any home-security footage. In multi-cat homes, there's probably trouble brewing—sibling rivalry or conflict, often the subtle, silent kind.

ADVANCED CAT WATCHING

Signs of a scaredy-cat

Fear is a healthy response to danger, but if your cat often displays some or all of the following behaviors, they may be living in a state of perpetual fear or anxiety—due to genetics, early life experiences, or stressful situations. If so, it's time to assess your cat's habitat and look at ways to help them feel at ease (see pages 46–47).

The look of dread

Anxious or fearful cats are all eyes and ears. They often blink more frequently and may stare with dilated pupils to keep threats visible while planning their next move. Some avert their gaze to appear nonthreatening; others force their eyes closed if they are avoiding detection. "Airplane ears" are flattened sideways and downward, and can be rotated independently to track scary noises.

Frozen with fear

Scared cats inhibit their movements to avoid being seen. They tense their muscles and crouch to appear smaller and to protect their vulnerable tummy. Their head and tail are drawn in close, and their paws remain grounded in preparation for running or hiding. It's unwise to assume a still cat isn't worried by what's going on, as being cornered or held prevents escape, which is very frustrating (see pages 102–103).

Hiding

Withdrawing from view makes a frightened cat feel safer from anything unfamiliar or intimidating. Tight spaces or high places make good hideouts, while inaccessible, dark, quiet spots help calm heightened senses. If your cat spends a lot of time in such places, it's a big clue they're uneasy about something in their world.

Hypervigilance

When all their senses are fired up, cats can be easily spooked and may dart around frantically at any unexpected sensory input. For some cats, living on the edge has become a way of life. Next time you assume your kitty is asleep, look closer—eyes forced shut with creasing corners might suggest they're still on alert.

"Problem" behaviors

Anxious cats can be frustrating to live with, but many of the behaviors that drive you crazy or have you pondering a vet visit—clawed carpets, urine-soaked duvets, and turds on the doormat—are clues that your cat's world isn't as relaxed as you think it is. Inadequate litter boxes, closed doors, loud music, nothing to scratch, and nowhere to escape from that tail-pulling toddler are all threats to your cat's happiness.

My cat seems to have a sixth sense

Somehow they know when it's time for the flea treatment; they run and hide as soon as the package comes out. It doesn't hurt them, so why all the drama?

the function?

A cat's instinctive need for control, reinforced by their sensitive nose and past negative associations, lead them to try to avoid any dreaded event.

What's my cat thinking?

With their acute observation skills, cats are masters at picking up cues that a dreaded event is about to happen—whether it's the application of their monthly flea treatment, a trip to the vet, or a vacuuming session. In the case of the former, your body language, along with the distinctive look and sound of the package and vial, are definitely bad omens that are likely to prompt your cat to take evasive action. The flea-treatment solution feels cold, stinks, and stings if it gets on or near a wound—and if your cat has ever unwittingly tasted it (inadvisable), it will no doubt have left a lasting impression on all their senses.

Inaction is still a reaction

We tend to focus on the active survival strategies for dealing with perceived threats—fight, flight, fidget, and faint—but scared cats often freeze. Never assume a cat is coping well with a stressful experience just because they don't protest. Hunkering down, hiding, and remaining still are all attempts to avoid detection and confrontation.

What should I do?

- **Prepare ahead** to make any unpleasant event as stress-free as possible for your cat. For instance, open the package and pop the top in another room, before you approach your cat—having already sussed out exactly where it needs to be applied. Vacuum when they are in another room, and in the case of a vet trip, see pages 150–151.
- **If you need to catch** or restrain your cat, stay calm, talk in a soft, soothing voice, and beware of sharp claws and teeth! A large towel and an extra pair of hands can make the process swifter and smoother.
- **Encourage your cat** to acquiesce by making positive associations with lots of fuss and treats.

> Don't be tempted to use flea products meant for dogs, or natural remedies such as garlic and tea tree oil, as they are toxic to cats.

SURVIVAL GUIDE

Moving to a new home

A disrupted routine, the loss of familiar comforts, and stressed-out owners are enough to send any cat into a tailspin! Thinking ahead is the key to making them feel more secure when you move.

1

Be prepared

A few weeks before the move, check that your cat's vaccinations are up-to-date and provide your vet, pet insurers, and microchip provider with your new contact details.

2

Promote calm

Preempt your cat's stress by using calming supplements or plug-in pheromone diffusers. These will be most effective if you start them a few weeks ahead of the big day.

3

Make a "safe room"

Set up a sanctuary with food, water, litter box, bedding, toys, and the carrier. Put a "DO NOT ENTER" sign on the door and keep your cat here, away from the chaos, for 24 hours before moving.

4

Replicate the calm

If you have access to the new home before you move in, create another sanctuary for your cat to go straight into. If not, keep them safe in the carrier while you go and set one up.

5

Settling in

Once you've unpacked, drape blankets, curtains, or bedding from your old home over any new furniture; leave them there for a couple of weeks so things smell familiar and homey.

6

Time to explore

Let them explore their new home at their own pace, allowing access at all times to the "safe room." Set up a treasure hunt using cat treats to entice and reward calm, curious exploration.

My cat is such a messy eater

They drop their food on the floor before eating it and make a mess all around their bowl and up the wall. It's like having a toddler!

the function?

Cats learn to lick delicate, precut mouthfuls of meat from a bowl, but their inner wildcat is urging them to grasp, shake, tear, and slice their prey's tissues.

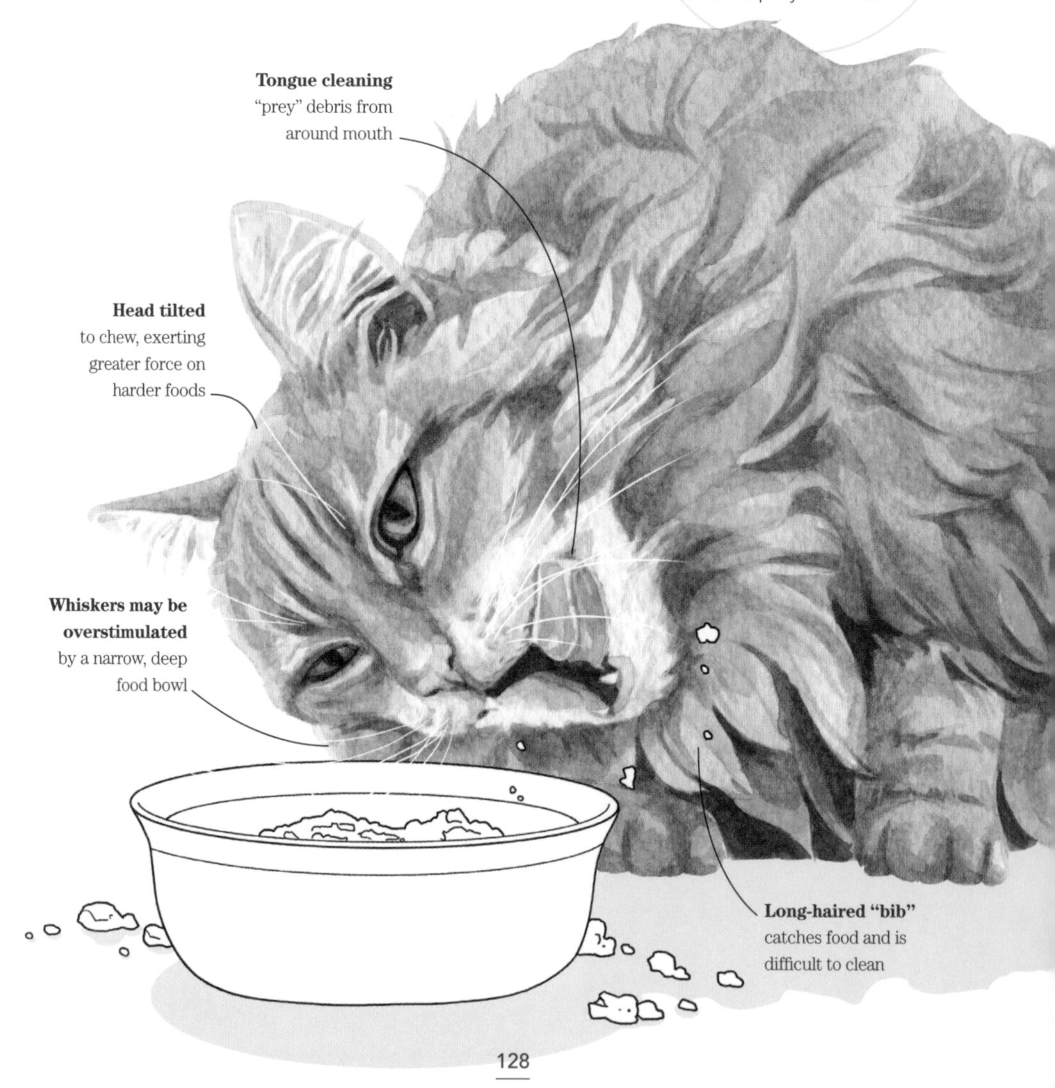

What's my cat thinking?

If your cat followed their natural instincts, they'd be wolfing down a mouse or plucking the feathers off a bird, not delicately picking meaty chunks from a porcelain bowl. There are no food bowls in the wild, so presenting a meal to a cat in one is more about what's easier for us.

Taste and touch should be center stage, but a dirty bowl, lingering soap residue on the bowl, or other factors (see below), could give a cat sensory overload and make them start thinking outside the bowl.

Sensory overload

A cat's whiskers are sensitive to touch while their ears are sensitive to noise (see pages 16–17), so the combination of a deep-sided bowl and a collar tag or bell could be irritating or uncomfortable when eating. Similarly, exposed nerve endings in teeth are unlikely to enjoy fridge-cold food or being continually knocked against a ceramic, glass, or metal bowl. It's little wonder some cats get frustrated and revert to nature's original food bowl—the floor.

What should I do?

In the moment:

- Let them get on with it and wipe up afterward.

In the longer term:

- **Watch your cat eat** on a regular basis so you get to know their norm, which makes it easier to detect early signs of pain. Trouble grasping or holding onto food, excessive or exaggerated tongue flicking, or head tilting and tooth grinding can indicate painful conditions in the mouth, spine, or digestive system.
- **Upgrade their dinner service** to a wide, shallow bowl or a feeding tray. Put a plastic mat underneath to make it easier to clean up.
- **Consider a silicone bowl**, rather than a traditional ceramic or metal one, as it's gentler on teeth, unbreakable, and dishwasher and microwave safe.
- **Avoid using heavily scented** detergents when washing their bowl and always rinse it thoroughly.
- **Repurpose an old bathmat**—some cats find it easier to eat dry treats or kibble from a rough or textured surface, and you can just throw it in the laundry every week.

My cat is a door dasher

My cat sits at the door ready to bolt. I worry they're going to run out into the street and get hurt, or just never come home.

the function?

Cats are naturally curious, and there's a great big world out there to explore. It's scary for some, but irresistible for others.

What's my cat thinking?

It's no secret that cats hate closed doors (see pages 118–119). Plus, the outdoors—or the hallway of your apartment building—is full of sounds, smells, and sights they've never encountered. If they go outside (see page 62–63), they may not understand that the front door is not as safe as the back door.

What should I do?

In the moment:

- **Keep toys and treats** next to the door. Toss one as far from the door as possible, just before you open it, so your cat runs the other way. They're then on hand to lure an escapee back in, too.
- **Put a doormat** with a texture your cat doesn't like to walk on right in front of the door. A clear plastic carpet mat, nubby side up, is ideal.
- **Block the access**—you may be able to put up a screen door or a walk-through pet gate in a hallway or vestibule leading to the front door.

In the longer term:

- **Fit your cat** with a reflective, quick-release collar with an ID tag.
- **Teach your cat to wait** in a specific spot away from the door. Set up a cozy elevated perch. Using the same cue word every time, entice your cat to hang out on it by offering treats, cuddles, brushing sessions, and even meals there. When they get the idea, open the door a crack and keep handing out the goodies as a reward for staying on the perch. Gradually make the exercise more challenging.

Emergency recall

Most cats come running whenever you feed them, so take advantage of that association. Call your cat's name before each mealtime. Repeat this in between meals by shaking the treat bag and rewarding them with a treat for coming over. Practice every day, especially when your cat is in another room or is distracted. Repeat this exercise until it's second nature.

“
If your cat makes a break for it, the best thing you can do is stay calm and note where they go.
”
Eyes alert, watching for signs you're going to open the door
Ears pricked, listening out for approaching footsteps
Sitting in front of the door, ready to spring into action

ADVANCED CAT WATCHING

How cats learn

Life is one big learning experience. Understanding how a cat learns enables us to alter their behavior or offer an alternative outlet for it, reducing stress and keeping them happy and healthy. It also helps us shape a kitten into a well-rounded feline family member (see pages 18–19) and literally teach old cats new tricks.

Living in the moment

Cats' natural curiosity, instincts, and senses help them gauge what's going on and register any change. As mindfulness experts, cats don't ruminate over the past or plan ahead.

Learning by association

When cats encounter an object, person, animal, or situation, their brains log the emotion it triggers and the outcome for future reference. When it happens again, they anticipate what will happen next–enabling them to avoid anything they perceive may endanger their survival, and embrace things that enhance it.

First impressions count

Sometimes lots of repeated experiences are needed and other times it just takes one (usually bad) experience to leave its mental mark. Negative associations are all too common, especially dread at the sight of the cat carrier. On the other hand, the sound of your car pulling up outside is likely to be a positive association indicating that a meal is on the way.

Teaching new associations

We can make sure kittens get off on the right paw when experiencing potentially stressful new things by introducing them gradually and rewarding curious, relaxed behavior. We can also help adult cats make positive associations by pairing something they currently find unpleasant (such as being brushed) with a new, good experience.

Positive reinforcement

Cats respond poorly to negative treatment, quickly associating you with bad things. Inciting fear and anxiety about your presence doesn't aid further learning, address the motivation for their behavior, or show them what you'd prefer them to do instead. Even a quick rant counts as punishment. The more effective approach is to focus on rewarding good behavior and ignoring the bad.

Benefits of training your cat

It's not only possible to train a cat (many of the techniques are the same as those used for training dogs and even toddlers), it's also fun and rewarding, providing mental stimulation for them and a chance to enrich your time together. Importantly, it can be used to reduce stress and modify unwanted behaviors—two things that actively wear away at the cat-human bond.

My cat scratches everything

You name it, they've scratched it: the sofa, wallpaper, and stair carpet—everything except the scratching post. Surely their claws are sharp enough by now!

the function?

Scratching keeps claws sharp and muscles stretched. It's also a visual clue and scent mark for other cats, which indicates their frequently visited routes.

What's my cat thinking?

It's completely natural for cats to scratch—it's an important stress reliever and stretches out the body after a catnap—but it's often what they choose to scratch that's the problem. Cats don't see expensive carpet or wallpaper, just perfectly placed textured surfaces to engage their claws and reaffirm their presence in their territory by depositing scent (see pages 14–15). If there is tension between cats, scratching can become more frequent in zones that are hot spots for conflict, such as stairs, doorways, entry/exit routes, or near sleeping and feeding areas.

Letting off steam

Address any causes of pent-up energy and emotion, such as unfulfilled hunting urges or conflict between cats. Tap in to your cat's wildcat rhythm by introducing routines and stimulating play and exercise (see pages 46–47 and 182–183). Faux pheromones may also help calm and reassure some cats and reduce anxiety and tension between them.

What should I do?

- **Don't hurl insults** or cushions! Cats need to scratch—it's a basic welfare thing.
- **Identify potential reasons** for your cat needing to let off "steam," both at that moment and in their world in general.
- **Provide scratching posts** or pads in your cat's preferred texture (sisal, rattan, cardboard, carpet, or wood) and angle (horizontal, vertical, or ski-slope). Ensure they're sturdy and tall enough to allow a full cat stretch (at least 35 in/90 cm). Place one in the scratching location of choice and one or more in hot spots.
- **Redirect your cat's scratching** onto the new posts and pads, with synthetic pheromones and catnip.
- **Deter "undesirable" scratching** by blocking access or using double-sided sticky tape on furnishings.

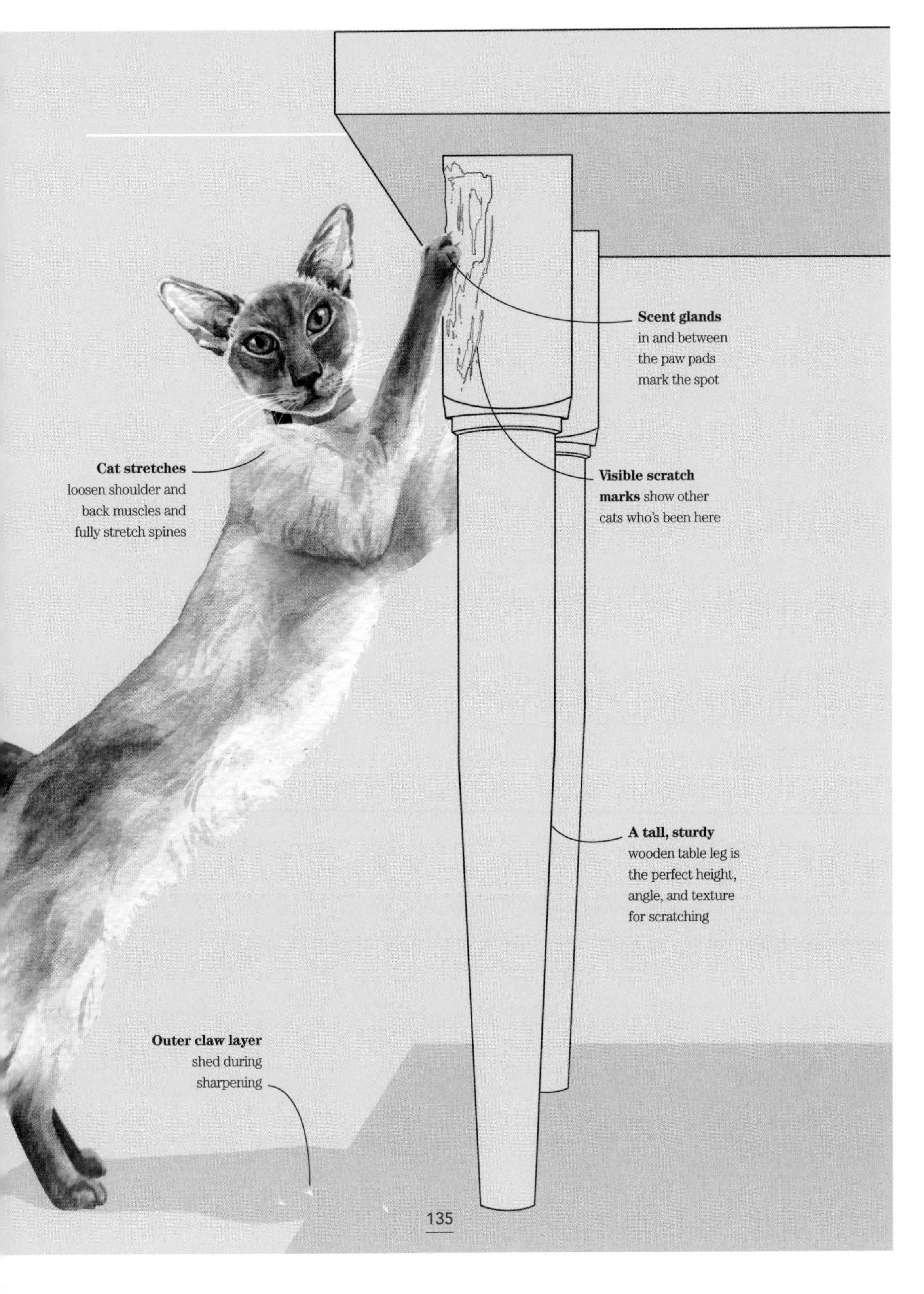
Scent glands in and between the paw pads mark the spot
Cat stretches loosen shoulder and back muscles and fully stretch spines
Visible scratch marks show other cats who's been here
A tall, sturdy wooden table leg is the perfect height, angle, and texture for scratching
Outer claw layer shed during sharpening

My cat is a counter surfer

They cruise the kitchen worktop like it's theirs. I've squirted them with water but now they just do it when I'm not looking. They've even figured out how to open the treat cabinet!

What's my cat thinking?

Your little cabinet-raider isn't scavenging, they're foraging. It's part of a cat's survival instinct to be always on the lookout for the source of their next meal. In the wild, that's a full-time job.

Your cat might be feeling hungry because of a disease or medication that has boosted their appetite, or perhaps they've just got a rumbling tummy (see pages 160–161). Either way, they've learned that where the kitchen's concerned, messy humans means free snacks could be on offer. Of course, your mini-explorer might just like the view from the worktop.

Paws for thought

Did you know most cats have a preferred paw? Males are more inclined to use their left paw and females their right. Persians are often ambidextrous, while more than 80 percent of Bengals are left-pawed. Which paw does your cabinet-opening kitty use? Set them some challenges and see which paw they lead with when playing, reaching for treats, or stepping over an obstacle.

What should I do?

In the moment:

- **Ditch the water spray**—bad human! At least, that's what your cat will be thinking. They'll just wait until you're out, as they'll associate the undesirable consequence with you, rather than with their actions.

In the longer term:

- **Understand the function** so you can redirect their behavior (see pages 32–33). If your cat is seeking safety or a view, introduce a stool or cat tower nearby; if the attraction is running tap water, a drinking fountain could be the solution.
- **Remove the rewards**—put away kitchen scraps and wash the dishes.
- **Make the worktop undesirable**—try placing a plastic carpet mat or car mat, prickly side up, on the surface. These deterrents tend to be effective because they're consistent and don't rely on you being present.
- **If boredom is the issue**, keep idle paws busy and engage your cat's inner forager (see pages 138–139).

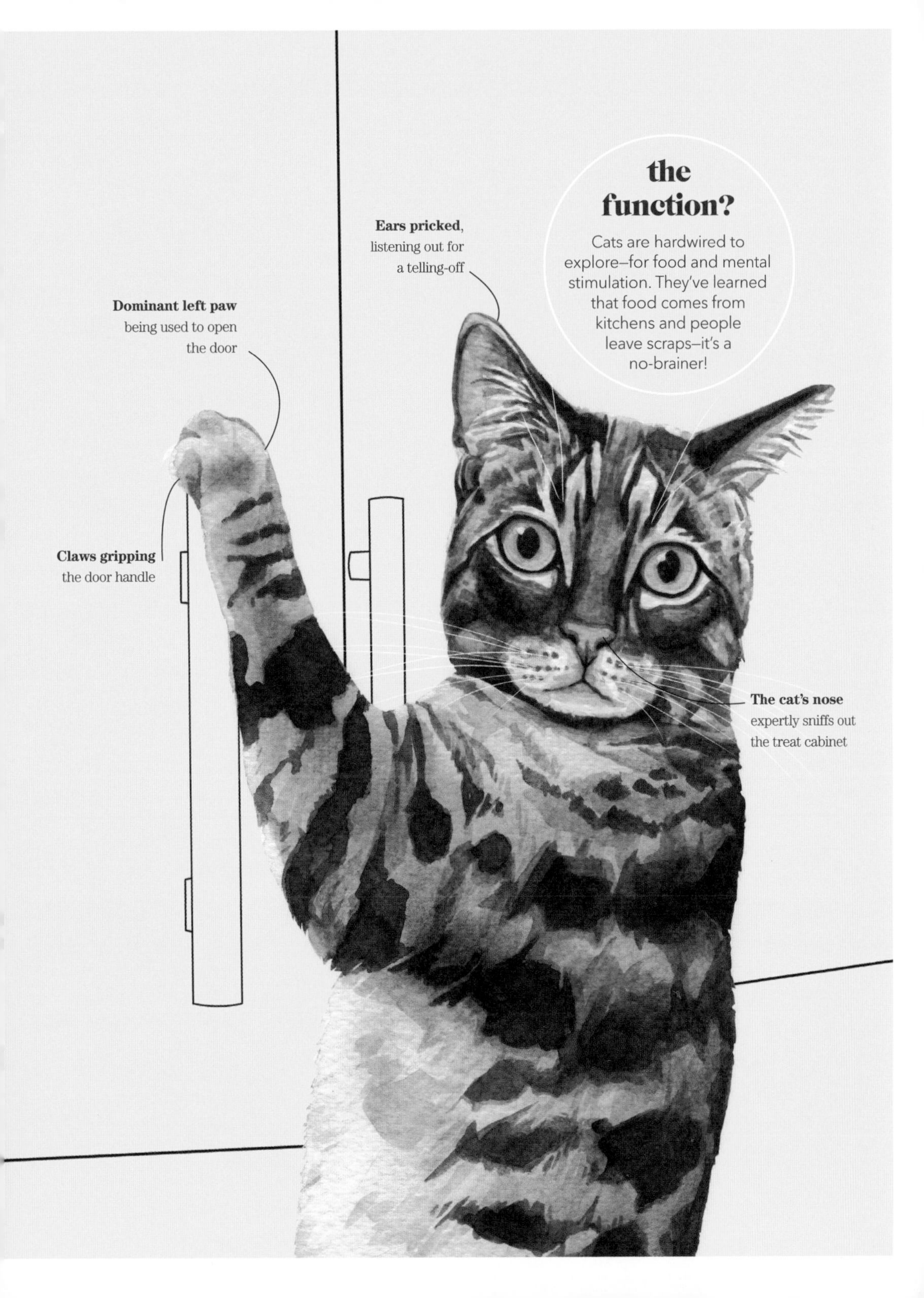

the function?
Cats are hardwired to explore—for food and mental stimulation. They've learned that food comes from kitchens and people leave scraps—it's a no-brainer!
Ears pricked, listening out for a telling-off
Dominant left paw being used to open the door
Claws gripping the door handle
The cat's nose expertly sniffs out the treat cabinet

SURVIVAL GUIDE

Food-foraging fun

Foraging for meals with toys and puzzles satisfies a cat's inner explorer. It's interactive, fun, and mentally and physically challenging, which is important for a cat's health and well-being.

1

Beginner puzzles

Try this for the first time when your cat is hungry. Beginners need easy wins, so half-fill a puzzle with food and scatter teaser tidbits around it. Keep adjustable openings wide to make it less challenging.

2

Wet or dry food

Extracting wet food from silicone lick mats or plastic mazes simulates tongue and jaw actions used when eating prey. Dry food is best in balls, rotating activity stations, snuffle mats, and refillable mouse toys.

3

DIY food fun

Household waste offers interactive possibilities. Cut holes in empty water bottles to make rolling treat dispensers, or create treasure hunts by hiding food in empty toilet paper rolls or yogurt containers.

4

Health benefits

Acquiring food via toys and games is a reward for problem-solving, persistence, and physical activity—just like hunting. It prevents boredom and reduces stress behaviors and illnesses. Puzzle feeding also slows down gobblers, reducing the risk of obesity and throwing up (see pages 160–161 and 166–167).

5

Old cats, new tricks

Puzzles enrich cats of all ages and abilities, but talk to your vet if your cat has underlying health conditions, as fun may turn to frustration if repetitive actions aggravate pain (see pages 146–147). Food puzzles may be difficult if disease (such as upper respiratory infections or nasal cancer) has affected appetite or smell.

CHAPTER FOUR

What's up with my cat?

Any seemingly normal or worrying behavior that stops, starts, or increases can be a sign of stress, pain, illness, or all of the above. Make sure these common cat complaints don't slip under your radar so you can seek timely advice from your vet.

My cat lingers in the litter box

They spend ages in their litter box, digging lots of holes, but there's nothing in there when they eventually emerge. Are they in need of a kitty laxative?

What's my cat thinking?

They're not hogging the litter box because they're checking their phone or engrossed in a good book; this is serious business and your cat needs your help. While constipation and diarrhea signal a vet visit, urinary problems are definitely an urgent concern. A sore bladder can quickly become a complete outflow blockage—in which case, your cat's thinking, "I can't pee!"—and that's an emergency. The stinging and constant urge to empty their bladder is painful and frustrating, as no amount of digging or squatting provides relief. Within hours, toxic kidney wastes and salts start building up in the bloodstream; if left, the bladder will eventually burst.

Spending ages answering a call of nature often starts as holding on too long because of pain, anxiety, or conflict.

Waterworks problems

Bladder inflammation in cats isn't fully understood, but there are lots of similarities with human interstitial cystitis. Faults with the bladder's protective lining and the brain–bladder nerve-pain pathways, together with anxiety, create the perfect storm. Being middle-aged, overweight, and male, and living indoors or with other cats are all risk factors. Stressful events or making sudden changes to their routine or home environment have an impact, too.

Persians are among the breeds that have an increased genetic risk of cystitis and bladder stones

Squatting low, assuming the usual urinating position

What should I do?

In the moment:

- **Contact your vet urgently**, even if you're not sure whether your cat is peeing and even if it's 2 am.
- **Check the litter box** to discover what's there—or not there. Is your cat passing huge pees or hardly any? Is the urine pink or bloody?
- **Collect any samples** you can find if there are accidents elsewhere—shower floors, sinks, and bathtubs are popular spots.

In the longer term:

- **Optimize litter-box management** (see pages 144–145)—urinary problems often start as avoidance of the litter box.
- **Reduce any causes** of anxiety or stress, and try to anticipate potential triggers in the future (see pages 144–145).
- **Minimize and phase in** any changes in your cat's habitat and routines—especially when it comes to their litter box, but generally, too.
- **Increase their fluid intake**—feed wet food instead of dry, with a little added water. Offer several drinking options—other than a bowl—in various locations (see pages 36–37).

My cat has stopped using the litter box

They've been using my duvet instead! I bought some scented litter to mask that horrible litter-box smell—could this be their idea of a pees-ful protest?

What's my cat thinking?

Rest assured, cats aren't spiteful or out for retribution—they don't think like that. If a vet check hasn't discovered any pain or illness, this behavior is probably down to anxiety or a negative litter-box experience. Even the most well-behaved cat won't hesitate to do their business around the home if something threatens their territory or sense of security—change is a major trigger. While you can't control bad weather or prowling local cats, you can create an indoor toileting experience that meets your cat's needs. It's easier to prevent this problem than to deal with it once your cat has developed a preference for soiling duvets or carpets.

What should I do?

In the moment:

- **Don't rant and rave**—more stress is the last thing your cat needs.
- **Clean up fast** so this doesn't become a habit (see pages 120–121).
- **Try to identify** your cat's motivation. Have you been scooping the box twice a day? Has anything changed within their territory or at its borders—windows, doors, cat flap, fences? Has access been restricted by a closed door, noisy washing machine, or another cat?

In the longer term:

- **Assess the location** of the litter boxes and, if necessary, relocate them to quieter, more secluded spaces—toileting is a vulnerable act, so cats need to feel safe and relaxed.

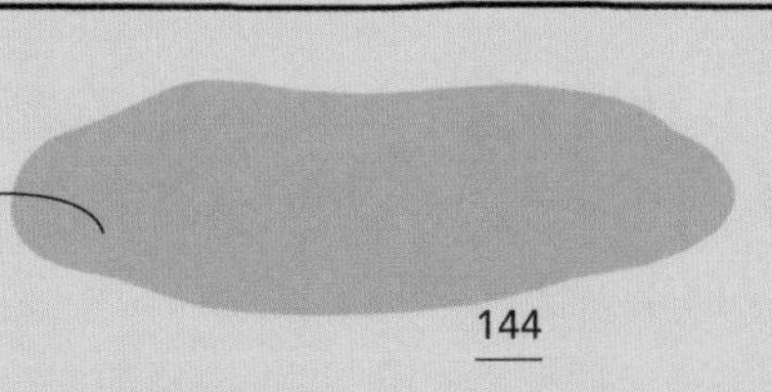

A pee patch on your duvet is a clear sign that all is not well in your cat's world

Paws are sensitive to textures—a soft duvet is more appealing than the new coarse cat litter

the function?

Using and sharing a litter box is a learned behavior that's not instinctive. It's your job to keep it appealing if you want to stop your cat thinking outside the box.

Wary expression, probably anticipating being yelled at

A bladder that needs emptying more frequently (and a litter box that gets dirty more quickly) is often a sign of illness

- **Offer options**—both open-topped and discreetly covered—so your cat can choose which they prefer.
- **Introduce changes gradually** and keep the old setup running until they're happily using the new one. See below for further guidelines.
- **Improve your litter-box etiquette**—one box per cat plus a spare is the rule, located in separate rooms.
- **Use a pheromone diffuser** to create a calming atmosphere, especially if your cat has been spooked while using the litter box, or ambushed by a cohabiting cat.

Litter-box laws

Do

- Buy the largest litter boxes you can accommodate—they should be one and a half times the length of the cat from nose to bottom
- Fill to a depth of 3–4 in (8–10 cm) with natural, clumping litter
- Position the litter boxes in quiet, private locations—cats don't like an audience any more than we do
- Scoop at least twice a day
- Empty and rinse the box weekly

Don't

- Use perfumed cleaning products, litter fresheners, or scented cat litter
- Over clean the empty box—a trace smell makes it familiar and reassuring
- Buy plastic litter-box liners—they catch in claws
- Fall for marketing gimmicks, such as kits for human toilet seats

ADVANCED CAT WATCHING

Signs of pain

Cats adopt a "keep calm and carry on" approach in the face of pain and ill health (see pages 164–165). Masters of disguise, they alter their behavior to avoid detection by predators and rivals, but this makes it tricky for us to spot the subtle or intermittent signs of pain that signal a vet visit is needed. Don't miss these common signs.

Altered mobility

Cats who are in pain tend to adjust how they move so they can stick to their routine and limit discomfort. They may walk or limp rather than run, and avoid heights or be hesitant, clumsy, or use intermediate surfaces when jumping. Toys, stairs, food and water, and litter boxes might all be a step too far. They might also be slower or stiffer when they get up or lie down; appear tense, hunched, or restless; or adopt unusual sleeping positions.

Withdrawing

When cats are in pain, unwell, or anxious, they take themselves off somewhere quiet, out of sight and reach of interfering people and other pets. It's an attempt to avoid further damage, get some rest, heal, and survive another day. This isn't a good long-term strategy, especially when there are vets and medications that can help.

Overattentive or inattentive

Sore teeth or joints can make eating and drinking, grooming, or maintaining claws unpleasant. Tail-down greetings, pulling away, or dipping down rather than rising into a pet, can be clues, too. You may notice repeated skin twitching or grooming over painful areas, such as the tummy (with cystitis), arthritic joints, swellings, or wounds—that's why hairballs can also be a sign of pain.

Vocalization

Silence might avoid drawing attention to yourself in the wild, but it's not so helpful if you're a pet depending on humans to help you. Cats won't meow, howl, grumble, or growl unless direct pressure is applied to a painful area by picking them up or stroking them. Cats in agony may groan, grunt, whimper, or even purr, but most suffer in silence.

Pained expression

A cat's facial expression can be a guide for vets assessing pain. Squeezed-shut eyelids and subtle changes in ear position and muzzle shape indicate that more pain relief is needed for cats recovering from trauma, surgery, or a painful illness.

My cat is so clingy and needy

They seem to know when I'm going out and then they supervise my every move and follow me like a shadow.

the function?

Your cat needs frequent, consistent opportunities to socialize with you–it's as vital for their happiness and stress-reduction as personal space and time alone.

What's my cat thinking?

Pet cats need human company, so it's normal for your kitty to be uneasy or frustrated when you go out for long periods. As far as they're concerned, when you leave, an interesting and pleasurable part of their life disappears. You know you'll return, but they don't. Certain social breeds, such as Bengals and Burmese, become particularly distressed when life takes you away from home. The daily ritual of curling up at your feet when you're working from home is reassuring, but when the routine changes and they're left alone, they can feel insecure and bored.

Cats are creatures of habit, so the unpredictable nature of modern life can make them feel out of control. Clinging to you means they won't miss clues that you're leaving again.

What should I do?

In the moment:

- **Don't fuss or feed** your cat as soon as you get up or arrive home, as it'll highlight your absence more.
- **Set up a food or catnip puzzle** when you get in, to occupy them while you get on with your chores.
- **Rule out illness**, which could be a reason for their clingy behavior, or be brought on by stress.

In the longer term:

- **Make time to play** and relax with your cat, ideally before mealtimes; keep to this routine on weekends.
- **Teach your cat** to have fun when you're not there with self-play opportunities (see pages 182–183).
- **Set up a sanctuary** in a room they like hanging out in, with a cozy bed in a sunny spot, above a vent, or on a heat pad. A bird feeder in the window, some online cat TV, or special cat music should keep them entertained while you're out.

Music for mournful kitties

Scientists have discovered that classical music, specifically slow string music, has a positive effect on felines' nerves. This theory was tested on cats undergoing anesthesia at the vet, when rock or pop music didn't hit the spot. There is even string music composed specially for cats, which can help reduce stress responses and could reassure them when they're home alone.

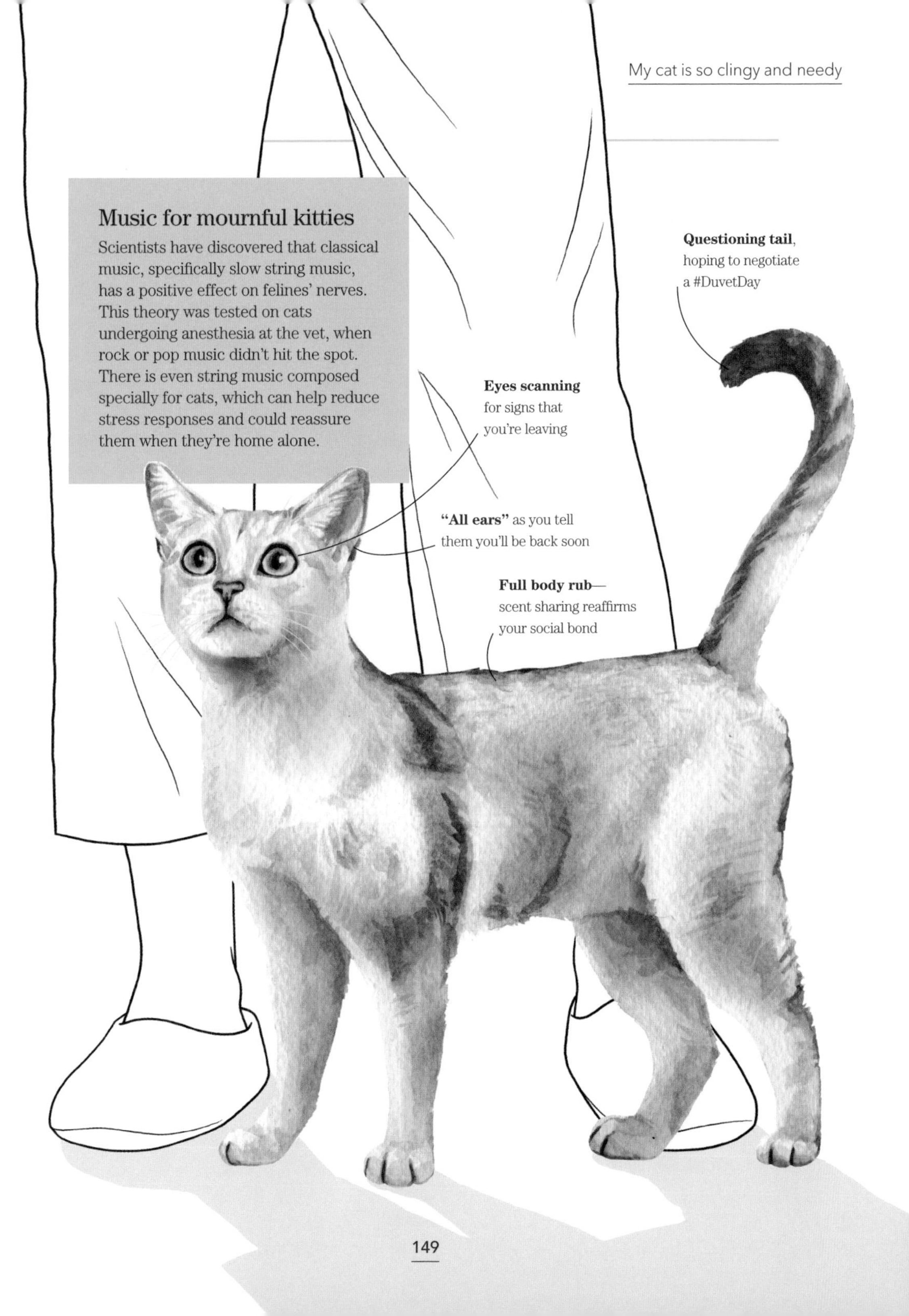

My cat hates going to the vet

They cry and pant all the way there. Then, once hauled out of the carrier, they behave perfectly for the examination but sulk and ignore me for days afterward.

the function?

Humans sometimes behave like predators, which shifts a cat's mindset from hunter to hunted. Any pursuit and capture can activate a rush of survival hormones.

What's my cat thinking?

Most cats hate going to the vet. It's a no-brainer—out of the blue, they're captured and caged, and taken from their safe, familiar home territory to the threatening climate of a vet clinic. It's the stuff of feline nightmares—an emotionally charged car ride followed by a sensory roller-coaster of strange sights, sounds, and smells from equally stressed cats and canine "predators." And that's before "Dr. Evil" prods and pokes them. On top of pain or illness, they've lost complete control, so they're anxious and fearful as well as frustrated and enraged. It's challenging for them on every level.

What should I do?

In the moment:

- **Avoid chasing and cornering** your cat—the pursuit will increase stress and activate "prey mode." Be covert and try to catch them unawares.
- **Stay calm** and talk in a soft, reassuring voice.
- **Tune in** to some soothing classical music en route (see pages 148–149).
- **Screen your cat** from worrying sights while traveling and waiting, by draping a towel over the carrier. This can double as a security blanket during the examination, providing warmth and familiar smells. It can mean the difference between your cat coping or not.

In the longer term:

- **Prepare ahead**—keep your cat in one room so you know where they are, get the carrier ready in advance, and allow plenty of time.
- **Find a gentle vet** who won't add to their stress (see pages 152–153).
- **Purchase a plastic carrier** with a removable top half that also allows you to give a reassuring one-finger cheek or chin rub when it's shut.
- **Help your cat** associate the carrier with positive experiences, such as toys and treats, and turn it into a cozy den.

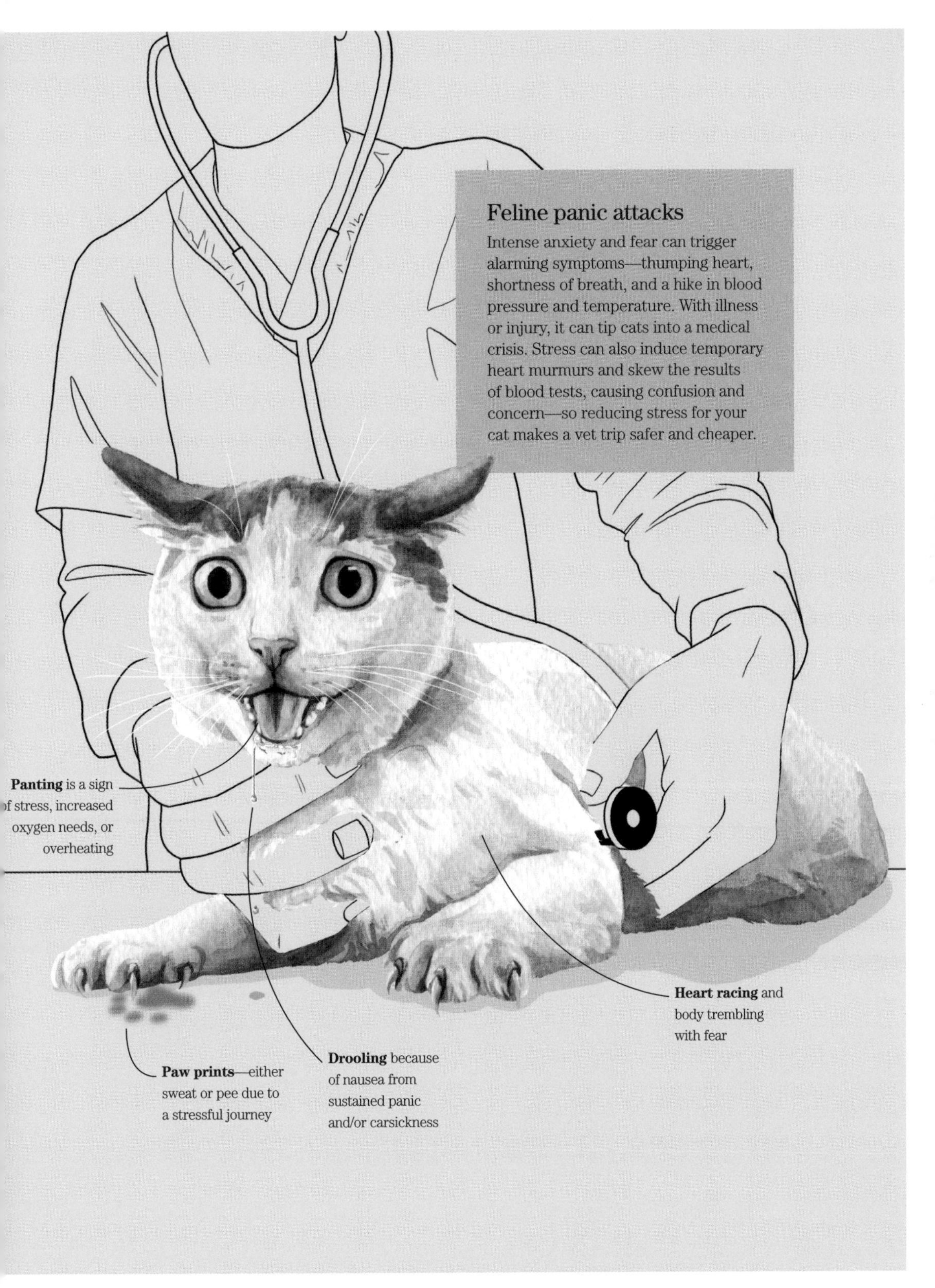

Feline panic attacks

Intense anxiety and fear can trigger alarming symptoms—thumping heart, shortness of breath, and a hike in blood pressure and temperature. With illness or injury, it can tip cats into a medical crisis. Stress can also induce temporary heart murmurs and skew the results of blood tests, causing confusion and concern—so reducing stress for your cat makes a vet trip safer and cheaper.

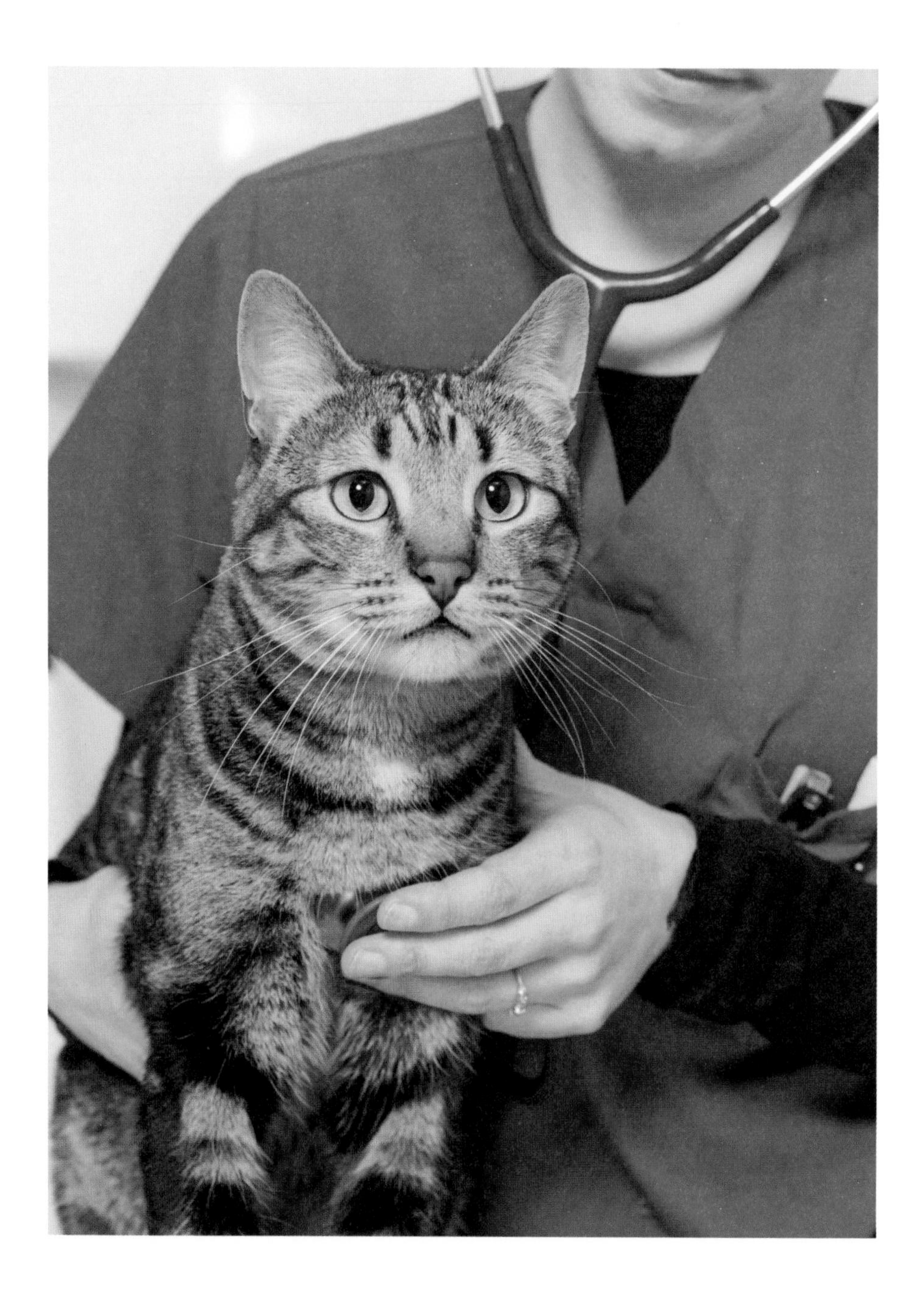

SURVIVAL GUIDE

Choosing the right vet

One of the most important people in your cat's life is their vet, so find one you trust to go the extra mile, to treat your cat like their own, and to look out for all aspects of their well-being.

1

A vet with "cattitude"

Look for a vet who takes time to listen to your concerns and treats your cat as the special individual you love. It's preferable to travel a little farther to see a vet who has a natural way with cats, rather than opting for the closest. Cat-only home-visiting vets are the ultimate way to keep stress levels low.

2

Special skills

Look for an AAFP certified Cat-Friendly Vet (see page 190) who continually refines their knowledge and skills and preferably works in a dog-free zone. Such vets are more likely to handle cats in the least stressful way and to recognize the link between physical and mental health.

3

Good vibes

Make sure the whole team is positive and working together, and the environment is calm, quiet, and hygienic. Book a tour without your cat and ask questions–most good vets are proud to explain how they work.

4

Trust your gut

Cats lash out when they're stressed, unwell, or injured. The right response from a vet is understanding and concern. Gripping their scruff or labeling them "the cat from hell" are signs of a vet who is behind the times.

5

Reputation and reviews

Always check out a vet practice's website, social media channels, and online reviews. Talk to friends, family, and neighbors, or ask for recommendations for a good local cat vet on your community forum.

My cat hates their best buddy

My two cats are littermates and they've never raised a paw to one another, until recently. They first started squabbling after one came back from the vet—could illness be affecting their relationship?

What's my cat thinking?

Subtle facial expressions, a glare, or a strategic sprawl near a doorway or on a stair landing might slip under our cat-conflict radar, but even best buddies disagree at times. Tensions can be heightened, too, when a cat feels vulnerable or starts acting differently because of pain or illness.

Smelling familiar to each other is important for cats. Changes in breath and urine—due to illness or medication—don't go unnoticed, while vet smells might trigger bad memories. A missing cat's group scent and territory markings fade, while remaining cats pine or take over their favorite spots.

Best friends forever?

Licking and rubbing between cats is usually a friendly gesture that cements bonds and creates a group scent (see pages 76–77). Other signs of good vibes are cheery greetings with upward tails and touching noses, and playing or snoozing together. But don't confuse an absence of scraps between cats sharing the same home and resources with being best buddies.

What should I do?

In the moment:

- **Don't pick them up** or leave them to sort it out among themselves. Act swiftly and try to pinpoint the root of the disagreement.
- **Distract them** without rewarding or startling them—shake the treat bag but don't feed them, or grab a cushion to block eye and physical contact.
- **Separate them** for 24–48 hours, each with their own resources, and watch for wounds and/or lethargy. Gradually reintroduce them as "new pets" (see pages 112–113) and monitor them for 48 hours.

In the longer term:

- **Be aware that** when a cat is away from home temporarily, the dynamics between remaining cat(s) readjust.
- **Preventing disputes** is kinder and easier than dealing with the fallout. Anticipate likely triggers, such as visits to the vet, and separate the cats afterward, as above, to allow them to groom off any unfamiliar scents.
- **Cats only form** social groups when there is enough space and resources

to go around. Help foster goodwill between cats by providing plenty of both (see pages 46–47 and 156–157).

- **Sharing scent rubs** and plug-in pheromone diffusers may provide reassurance and offset shifts in dynamics (see pages 14–15).

the function?

Feline "friendships" can be fickle—they're not essential for survival and cats prefer not to share, so they haven't evolved the skills to reconcile any differences.

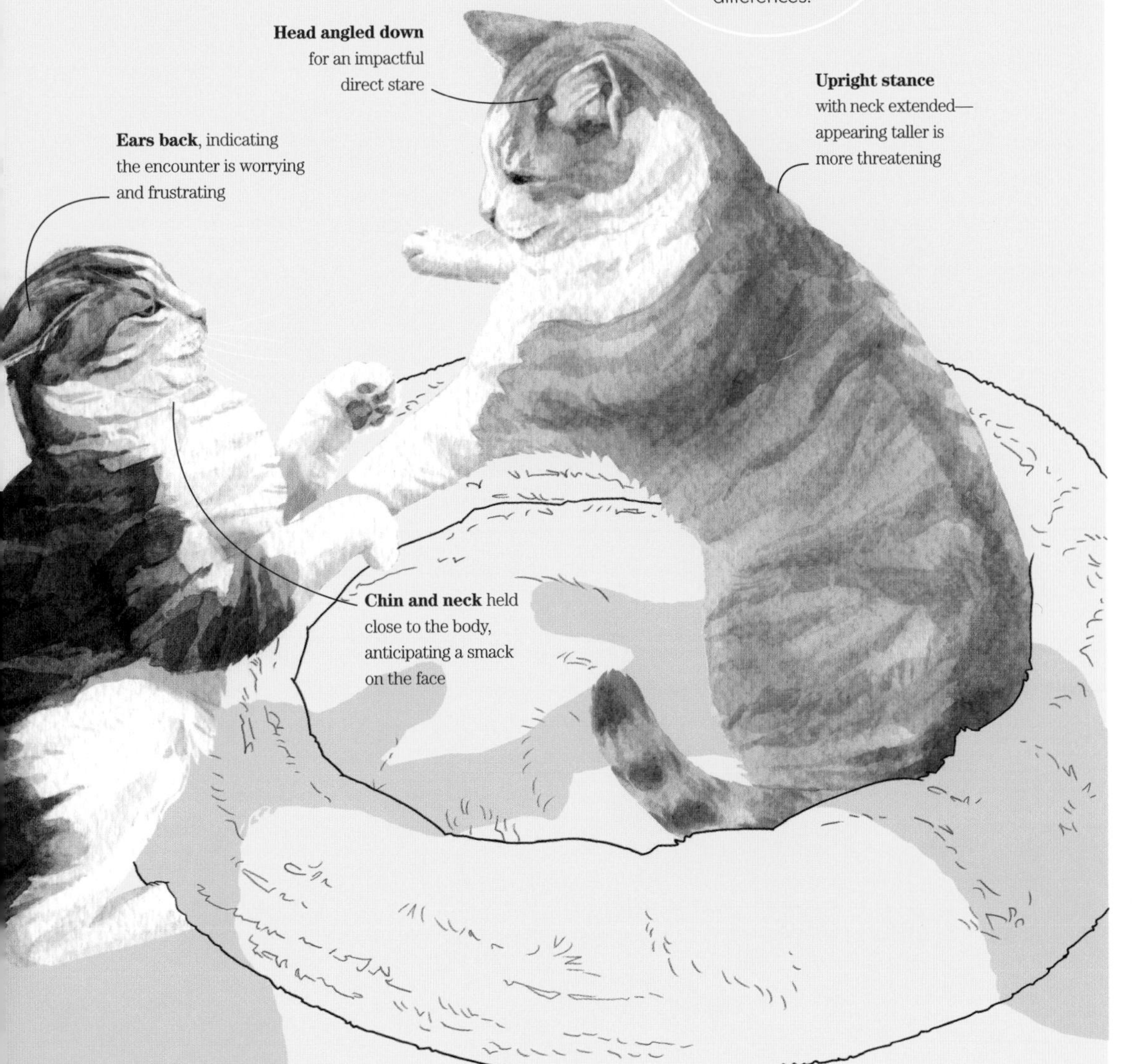

SURVIVAL GUIDE

Multi-cat harmony

Some cats thrive on being an only companion, while others seem happier with a buddy. Achieving a friendly group can be tricky, so check in to see if your cats are really #HappyTogether.

1

Feline the love

Multi-cat harmony isn't just the absence of physical violence. Look for friendly greetings with upright tails, nose touches, and face or body rubs. Playing, cuddling up together, or washing each other all suggest positive vibes, too.

2

Turf wars

Vocal or physical threats and armed attacks are obvious signs of conflict, but it's also not great if you never see any positive interactions. Look for silent threats, such as direct stares or strategic blocking of another cat's free movement or access to resources.

3

Plenty to go around

Life's a competition, and limited space or resources make the stakes higher. Your cats' habitat (see pages 46–47) has to work harder with each extra cat, so offer individual and separated resources for each, plus one extra for good measure.

4

Pay for good advice

Leaving fighting cats to sort it out themselves is a recipe for disaster. Ask your vet to check for pain or illness, to offer advice, and to recommend a good cat behaviorist (see page 190). Little things can make a big difference, so it's worth the effort.

5

Feline the stress

Sometimes signs of trouble in the fold are less obvious, such as scaredy-cat behavior (see pages 122–123); avoiding being in the same room; illness (see pages 164-165); or scratching, spraying, or litter box issues.

My cat is a picky eater

One day they like their food, the next they turn their nose up at it. I only buy the best cat food, so why are they so picky?

What's my cat thinking?

There's far more to a dining experience than the food. Imagine being in a noisy, busy restaurant, where the diners at the next table continually eyeball you and swipe bits of food off your plate. Nothing would kill the mood faster! And this can certainly be a source of anxiety for a cat living with other pets (see pages 110–111). Continuing the restaurant analogy, those of us who find comfort in familiarity will often reserve the same table and make the same menu choice every time, while others are eager to try something new. We all have our food preferences and comfort zones, and cats are no different.

A bitter pill

Studies show that feline taste buds detect amino acids (protein building blocks), bitterness, and saltiness, but not sweetness. Cats are impressionable—a bad experience can put them off eating the same food again, so bitter-tasting pills and potions shouldn't be hidden in food. Ask your vet about ways to disguise them using chewy treats, or whether they come in a more palatable form.

What should I do?

- **Rule out any underlying** medical reason for even a mild change in appetite—pain, nausea, illness, and stress may initially look like pickiness, so get a vet check.
- **Offer small quantities** more frequently—cats are built to eat small portions, so walking away after a few mouthfuls isn't fussiness. This also avoids food spoiling and attracting flies. Use a motion- or microchip-activated sealed bowl.
- **Don't force your cat** to eat foods they don't like. This may be individual taste, but some cats become phobic of new foods as kittens (see pages 18–19). Others crave diversity, which could be nature's way of maximizing nutrient balance or avoiding toxin or parasite buildup.
- **Offering morsels of new food** on the side of a plate, alongside familiar food, can make it less daunting.
- **Don't serve fridge-cold food**—room temperature or "prey temperature" (98.6°F/37°C) is best.
- **Consider the effects** of your cat's dining setup and psychological stress (see pages 128–129 and 166–167).

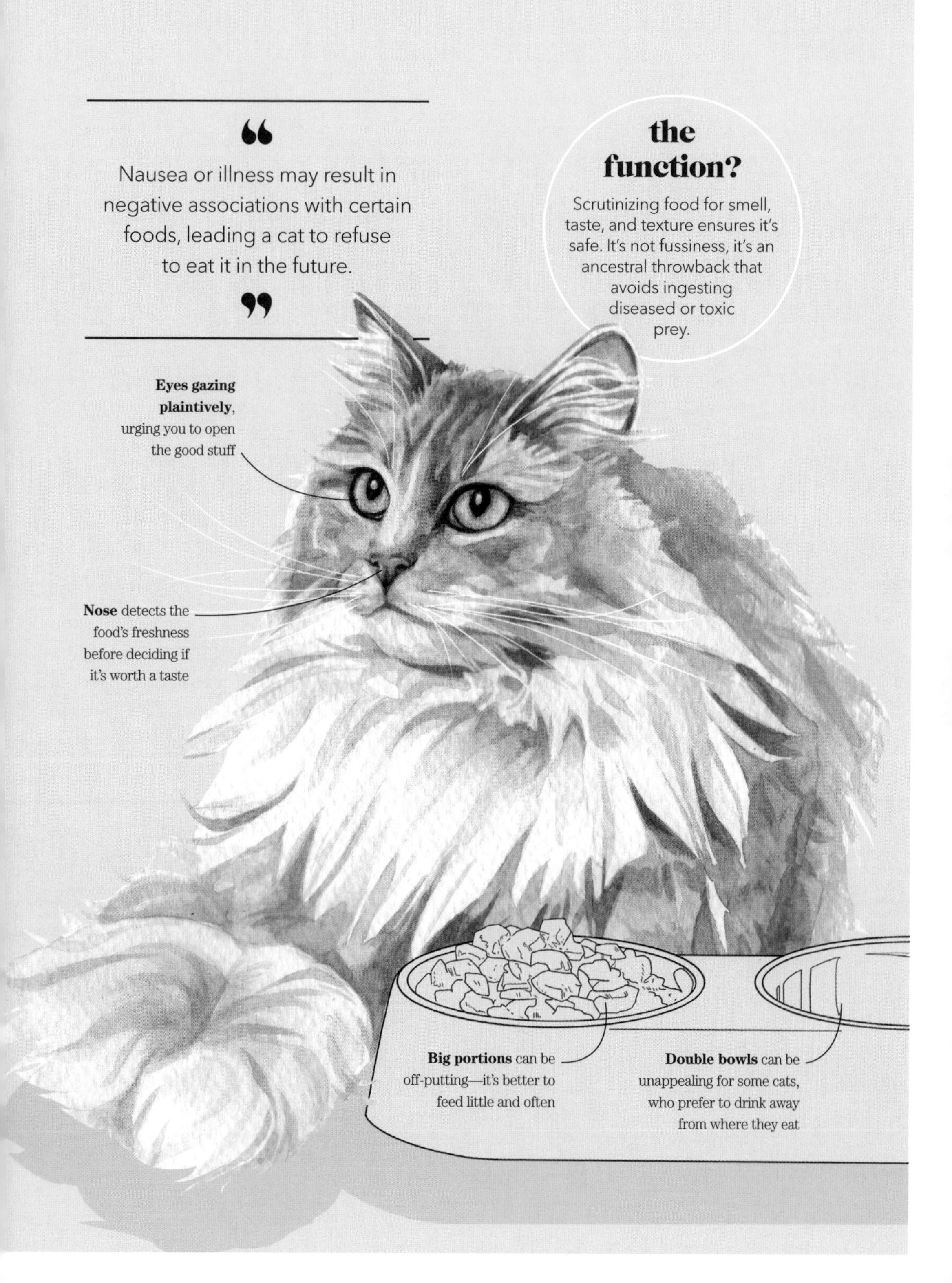

“Nausea or illness may result in negative associations with certain foods, leading a cat to refuse to eat it in the future.”
the function?
Scrutinizing food for smell, taste, and texture ensures it's safe. It's not fussiness, it's an ancestral throwback that avoids ingesting diseased or toxic prey.
Eyes gazing plaintively, urging you to open the good stuff
Nose detects the food's freshness before deciding if it's worth a taste
Big portions can be off-putting—it's better to feed little and often
Double bowls can be unappealing for some cats, who prefer to drink away from where they eat

My cat has gotten fat

The vet warned me my cat's overweight and at risk of early onset arthritis and diabetes. I insisted they're just big boned, but I'm wondering if I should have listened more carefully.

What's my cat thinking?

Tubby torties and plus-size pedigrees are destined for a plethora of health problems and a shorter lifespan. It's also a frustrating existence for a cat when normal feline activities, such as jumping, grooming, or playing, become physically challenging for them.

You may think that today's pet cats have it easy, dining on the plentiful and tasty ready-prepared meals we serve them. Now that we discourage their main form of employment and exercise, though, 50 percent of cats are overweight—and not happier for it.

Cats are opportunistic feeders and will generally eat more food than they need; anxious, lonely, or bored cats may comfort eat or gobble food (see pages 166–167). Humans express love through the sharing of food and tasty tidbits—it's a dangerous combination.

How does your cat shape up?
The ideal weight for an average-size cat is 8–10 lb (3.5–4.5 kg) . Vets use a body condition score (BCS) system that combines a cat's profile body shape with the palpable fat coverage over the bones of the ribs and spine.

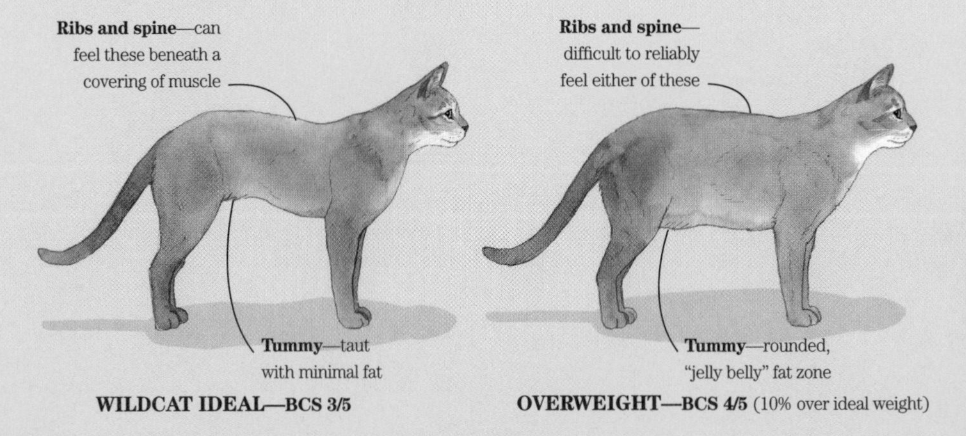

What should I do?

- **Get advice from your vet** before changing or reducing your cat's food. Some prescription diets are better for certain diseases, while cats with fatty livers can die if their calorie intake drops too quickly.
- **Stick to complete**, balanced foods, rich in moisture and protein, and add a few teaspoons of water.
- **Limit dry food**, which is calorie-dense, promotes dehydration, and usually contains less meat or fish.
- **Measure your cat's food**, sticking to the lower end of the manufacturer's feeding guidelines.
- **Divide their daily food** ration into multiple small meals to provide throughout the day, rather than one all-you-can-eat buffet.
- **Treats and tidbits** should make up no more than 10 percent of your cat's daily calorie count.
- **Encourage foraging**, which burns calories, promotes slower eating, and is enriching (see pages 138–139).
- **Play busts boredom and stress**, which are both causes of inactivity and obesity—and it makes burning calories fun (see pages 182–183).

the function?

There are no fat wildcats–hunting burns calories. Fat cats don't feed themselves; it's up to us to regulate their food intake and activities, in order to prevent illness.

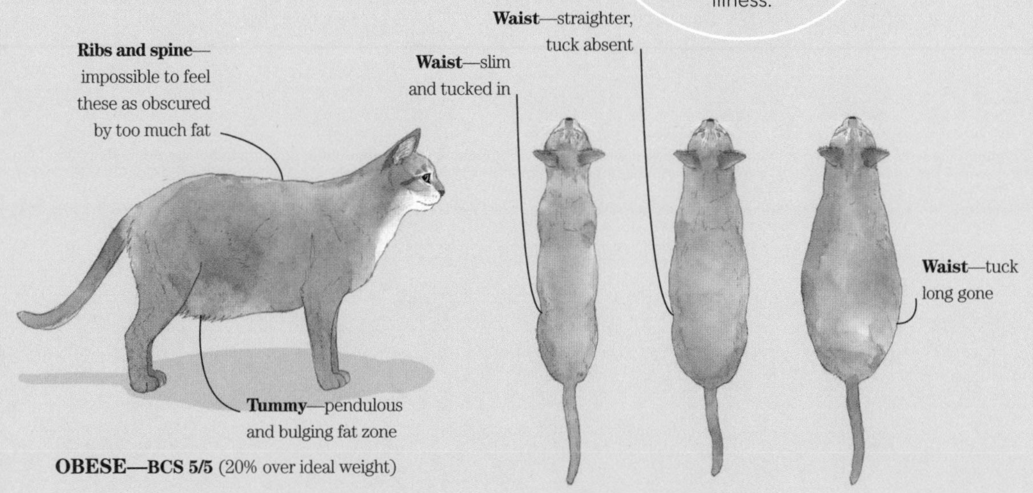

OBESE—BCS 5/5 (20% over ideal weight)

My cat seems bored

They've got baskets full of toys they never play with, and look at me with disinterest when I pull out the fuzzy mice on a string. What can I do spice up my cat's life?

the function?

In the wild, cats might hunt ten times a day or more. When there's no need to hunt, they need something else to do.

What's my cat thinking?

Feline play is always a form of hunting. Real prey is clever and unpredictable. It runs at different speeds. It changes direction. It scurries under the couch or behind the curtains. It plays dead and then suddenly jumps up and makes a break for it. Interactive play with us is the only thing that can simulate that experience. But old toys are boring—"I've killed that mouse a thousand times already!" And we need to tap into the stalk and pounce sequence (see pages 70–71), rather than expecting the cat to leap around like mad. From the cat's point of view, leaping is a last-ditch attempt at a catch. If your cat is jumping for a toy, they're not hunting.

No-tech toys

All those wind-up, hang-on-the-door, and automatic motion-detector toys don't act like prey. Most move in a simple pattern. But cats are extremely intelligent hunters and can quickly figure out these patterns. And then the toys are just no fun.

What should I do?

- **Play like prey**—slowly move the toy under a cushion or behind you and out of sight, and your cat will be dying to find out where it went.
- **Watching and planning** are part of the hunt, so remember that a cat who is not moving but is locked on visually to a toy is still engaged in the game.
- **Rotate the prey**—offer a few toys to hunt in each session and pack them away in between play times. Studies show that cats get bored of toys if they're left around—novelty is important.
- **Create opportunities** for foraging (see pages 138–139). Fold up a towel, hide some treats among the folds, and put it inside a box or a big paper bag. Fill a big delivery box with crumpled-up paper from your recycling bin. Then toss in small toys and treats so your cat can go dumpster diving.

Nature did not design such an efficient predator to lie around all day doing nothing.

ADVANCED CAT WATCHING

Signs of illness

Long-term fear, anxiety, frustration, or pain can wear down the body and mind. Cats living with animals they dislike or missing something they need from you or their habitat are more likely to become distressed and unwell. Spotting the early signs of stress-related conditions means you can seek help before things escalate.

Immune system defects

Stress hormones inhibit a cat's immune system, making them more susceptible to infections and increasing the risk of cancer; allergies; and inflammatory conditions of the gut, urinary tract, and skin. Symptoms of these illnesses vary, but most result in a picky appetite, weight loss, and reduced energy levels, while infections can also cause a fever.

Urinary tract issues

A stressed-out cat's brain sends signals to the bladder that exacerbate inflammation and pain, and may result in cystitis. Anxious cats hide more and avoid going to the water bowl or litter box. The resulting stagnant, dehydrated urine irritates and stretches the bladder. Inflammation, crystals, stones, and blood can cause painful, fatal blockages to urine outflow (see pages 142–143).

Gastrointestinal issues

Long-term stress increases the production of stomach acid and impairs the lining of the gut. It interferes with normal gut blood flow and movement, as with irritable bowel syndrome, and disrupts the balance of helpful gut bacteria. This contributes to a more fragile lining of the stomach and intestines, making it more prone to irritation and ulceration. You might see signs such as vomiting or intermittent diarrhea, which can get worse with certain types of food and may lead to weight loss. Gut pain and nausea can also blunt a cat's appetite.

Skin conditions

Some stressed cats lick and scratch themselves repetitively as a kind of nervous habit. Others have allergies to food, pollens, dust mites, and so on, brought on by long-term stress on the immune system. Any disease causing intense, ongoing itchiness is wearing and stressful in itself. It's complicated, but look out for skin lesions–scabs, crusts, blisters, pustules, or sores.

Top tip: Choose a vet who treats your cat as a whole, and who realizes that making them feel better doesn't just involve prescribing medication, but improving their habitat and mental health, too (see pages 152–153).

My cat speed-eats and throws up

They eat so fast, they practically inhale their food, and then it all comes back up again a few minutes later, usually on the rug. Sometimes they even eat it again—it's so gross!

What's my cat thinking?

When recently guzzled food suddenly makes a reappearance it's tempting to think your cat is greedy, but this is often a sign of anxious speed-eating. Gobbling can be a habit learned as a kitten, or from falling on hard times and going hungry. Sometimes it's because of tension between cats at mealtimes, or some other anxiety about the world they live in (your home). Unpredictable or infrequent feeding can make a cat frustrated and hungry, so when they're presented with food, they wolf it down. Having not even hit their stomach, it comes back up undigested, looking, smelling, and—yes—tasting good enough to eat.

What should I do?

In the moment:

- **Before cleaning it up**, check it for color, texture, and anything that shouldn't be there, such as fur, plastic, or string. Note when your cat last ate anything, and what happened just before and after. This will help your vet decide which is more likely, stress gobbling or an underlying illness. It sounds nasty, but it can be really helpful for your vet to see photos of what's come up.
- **Offer more food**, but only a tablespoon at a time and spaced out over the course of an hour. If the same thing happens again after eating slowly, it could be due to an obstruction, so call the vet.

In the longer term:

- **Get a vet check**—bolting food down too quickly can be a sign of pain (often in the mouth or neck) or of certain illnesses. Discuss whether a change of diet might help.
- **Use a puzzle feeder** to slow down their eating (see pages 138–139). Manage their hunger and frustration by serving small, frequent meals.
- **Reduce any anxiety** around eating (see pages 158–159). Put food and water bowls well away from litter boxes. Feed multiple cats in different rooms. Some cats also feel safer eating above floor level.
- **Nip it in the bud**—stress causes digestive diseases and vomiting, so it becomes a cycle if it's not resolved.

> Regurgitating undigested food is different from vomiting stomach contents, but both can happen with stress or illness. Your vet will be able to decipher the difference.

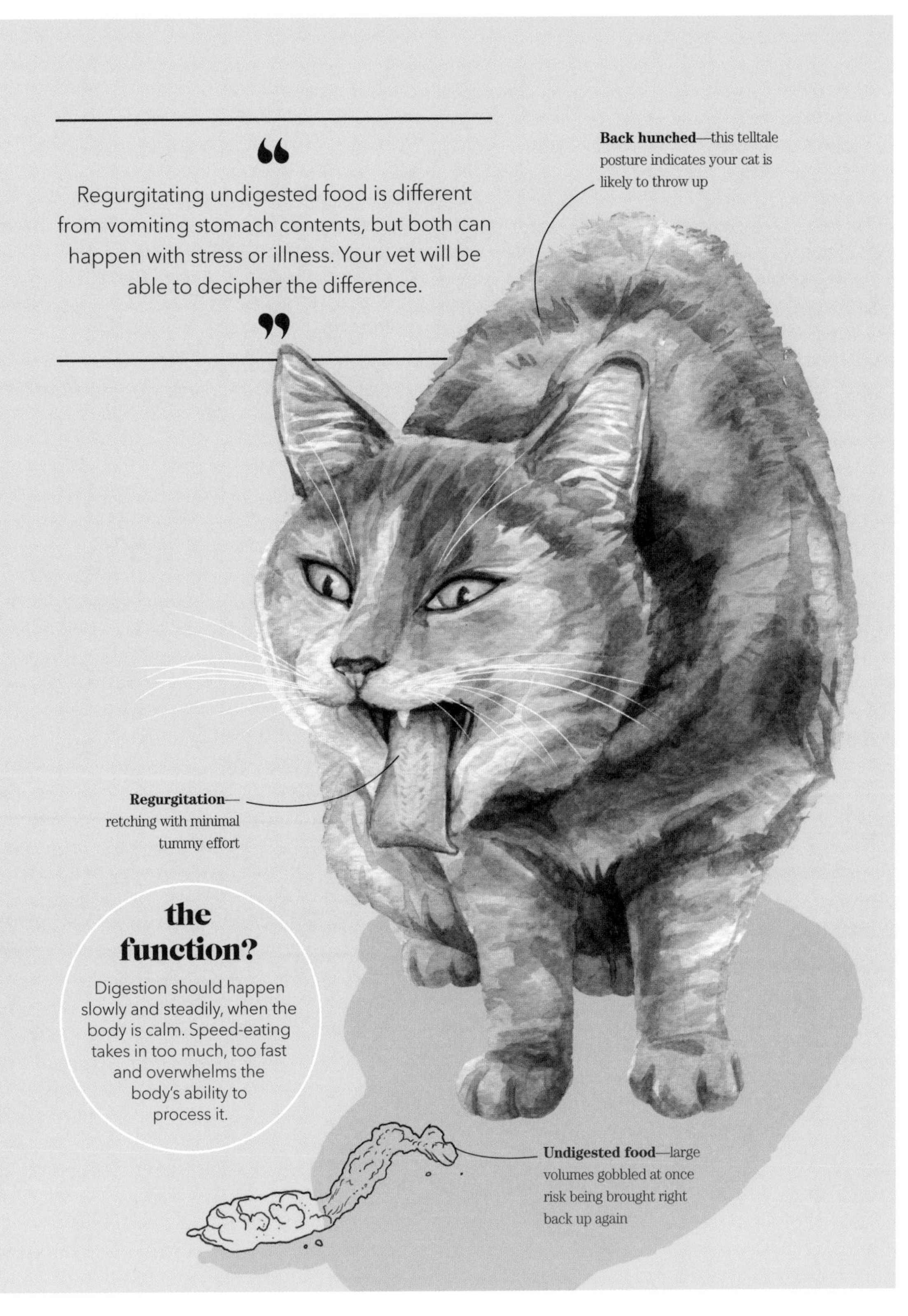

the function?

Digestion should happen slowly and steadily, when the body is calm. Speed-eating takes in too much, too fast and overwhelms the body's ability to process it.

My cat is unpredictable with visitors

When cat-loving friends visit, they're touchy or aloof, but they're all over my uninterested plumber. What's that about?

What's my cat thinking?

You're probably not always in the mood for entertaining guests and neither is your cat. Everyone who visits the house is different—and cats don't just see these peculiarities, they smell, hear, and feel them, too.

A person's nonverbal cues can also have a significant effect on their approachability. People who don't have an affiliation with cats come untainted with the aroma of other kitties. They are also less likely to try to engage or make eye contact with your cat, which makes your cat feel more relaxed and allows their natural curiosity to take over. In the end, it's your cat's prerogative to decide who earns the title of "cat person."

What should I do?

In the moment:

- **Brief guests to ignore** your cat. This will give your kitty time to make up their own mind.
- **Line visitors' pockets** with your cat's favorite edible bribes to pique their interest.
- **Supervise interactions**—if your cat isn't relishing the attention or is showing signs of feeling grumpy (see pages 102–103), put a stop to it and distract them before anyone is hurt.
- **Ensure your cat** has an easy exit route. Giving them that choice and control may avoid them lashing out.
- **Set up a puzzle feeder** elsewhere in the house to keep your cat busy, so they can adapt to visitor noises without seeing or smelling them.
- **Visitors will smell** funny to your cat, so ask them to wash their hands with your familiar soap when they arrive, and put their bags and shoes away.

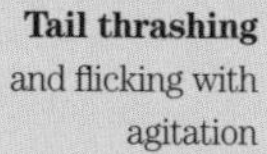

Tail thrashing and flicking with agitation

Guest rules of engagement

1. **Wait to be invited.** Only fools rush in—always let a cat approach you first.
2. **Be empathic.** Respect the cat's needs and don't impose your own.
3. **Read the signs.** A cat's vocal clues and subtle body language will either be encouraging more attention or asking for more space, so pay attention.
4. **Quit while you're ahead.** All good relationships take time, so keep interactions short and don't expect too much too soon.

the function?

As territorial prey animals, cats are naturally wary of uninvited guests, especially feline fanatics who reek of other cats, which they're programmed to avoid.

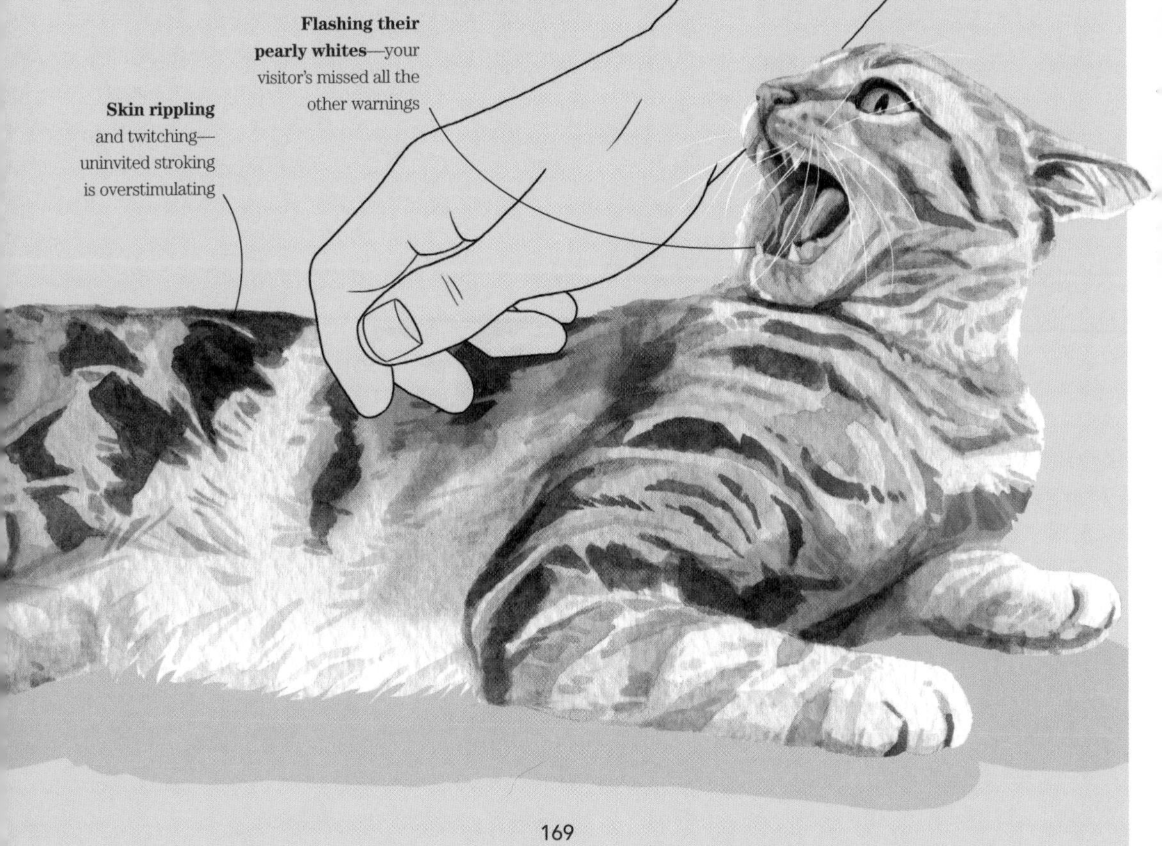

Flashing their pearly whites—your visitor's missed all the other warnings

Skin rippling and twitching—uninvited stroking is overstimulating

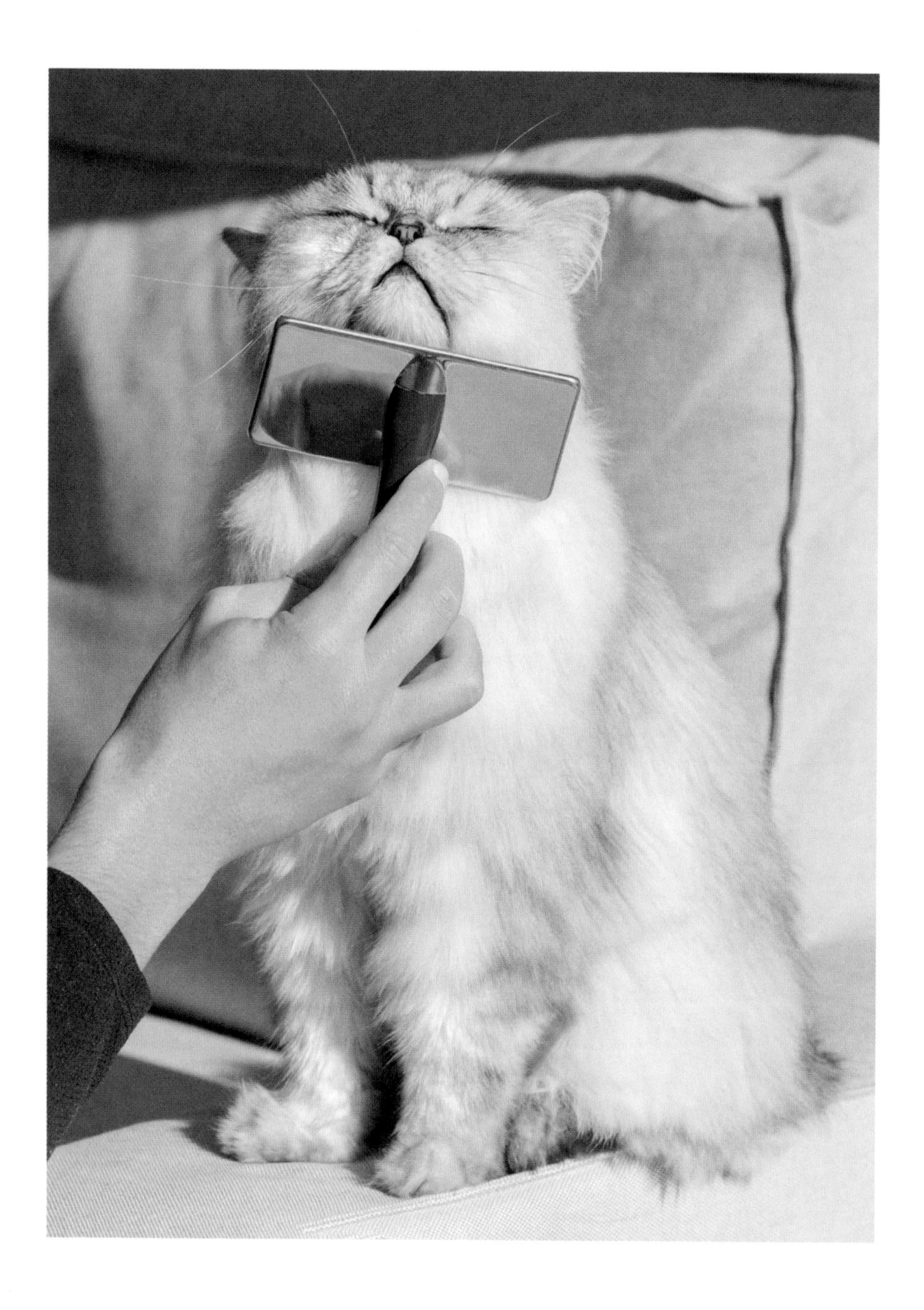

SURVIVAL GUIDE

Good grooming

Happy, healthy cats keep their fur impeccably groomed, but if they're below par, have dense or long coats, or are shedding more than usual, they may need some help to prevent knots and mats.

1

DIY grooming kit

Only use a soft slicker brush, a brush with a soft tip, or a silicone brush, as they are all gentle and effective at coat maintenance. Gently tease knots apart with your fingers and avoid scissors, as it's very easy to cut the skin. After brushing, sweep damp hands over your cat to remove loose fur and prevent post-grooming hairballs.

2

Gently does it

First, train your kitten or cat to enjoy daily massages and strokes when they're calm and relaxed. Then use edible rewards to create positive interactions with grooming. Avoid tugging by working on the coat in small sections. Short, positive sessions are best–don't push their patience too far.

3

Health fur-ometer

Poor coat condition can signal anxiety, obesity, malnutrition, or other underlying health problems, and matted fur is uncomfortable. Cats who seem cantankerous when groomed are likely to be fearful, unwell, in pain, elderly, or all of the above. Book a vet check and mention that matted fur may need to be removed.

4

Nailing it

Indoor or elderly cats may benefit from regular claw trims, so ask your vet to show you how. Use positive associations to encourage your cat to allow frequent paw inspections for signs of nail damage, thickening, or overgrowth. Eventually introduce the nail clippers, but be aware that arthritis and overgrown nails are very painful.

My cat is a clean freak

I've known cats to stop washing when they're ill, but my cat has the opposite problem. Is something wrong or are they just fastidious?

What's my cat thinking?

Washing is a normal behavior that keeps cats clean and well-groomed. It typically takes up half of their awake time, and they are such meticulous self-cleaners that we often don't notice external parasites, such as fleas, until either they're plagued with the critters or their overenthusiastic tongues have damaged their fur or skin.

Cats can groom excessively because of medical conditions that cause skin itchiness (such as allergies), unusual nerve sensations, and pain. Some cats develop skin lesions purely from stress and will then lick those areas more.

Over-grooming can also stem from frustration or a self-soothing response to conflict or anxiety, and is a symptom of feline compulsive disorders. It's complicated, so talk to your vet.

What should I do?

In the moment:

- **Check their body language**—is this just normal post-meal self-care, or could it be a response to pain or distress? Has your cat been agitated, using the litter box more frequently, or had a spat with another cat?

The multitasking tongue

A cat's barbed tongue serves many functions. It's vital for eating prey, enabling the cat to strip the flesh from the bones—and essential for washing up afterward. Its rough surface cleans away wound debris, dislodges parasites, relieves itches, and combs through the fur, while saliva has healing properties and cools the skin. A cat's tongue can be thought of as nature's own wound cleanser, but it's a balancing act between healing, infecting, and damaging. Thank goodness for antibiotics!

- **Note the target areas**—over-grooming harder-to-reach places tends to be because of pain or illness. Excessive washing of the tummy or private parts are clues not to be missed—cystitis in male cats can be life-threatening.

In the longer term:

- **Keep your cat's life** calm and predictable. Address any stressful situations, such as lifestyle or environmental changes and conflict with other cats (see pages 156–157).

the function?

Excessive grooming can be a cat's way of addressing a specific issue–whether that's stress, conflict, pain, or illness, you need to get to the bottom of it.

My cat is all itchy and scratchy

When they get a persistent itch, they scratch and nibble, and really get their teeth into the base of their tail, sometimes tugging out fur. They seem fine—it's probably just fleas.

the function?

Wildcats remove parasites—the most likely cause of itching—by nibbling with incisor teeth, scratching with back claws, or rubbing against rough surfaces.

What's my cat thinking?

Scratching helps relieve itching, but it can also happen in unexpected scenarios, such as altercations. Instead of chasing off another cat or running away, your cat may start to scratch, nibble, or lick themself. Fidgeting in this way is the kitty equivalent of nail-biting and may distract from the conflict, or deflect it and reduce tension.

If you're noticing more than the occasional scratch, there could be ongoing conflicts, stress, or illness—including parasites such as fleas. If your cat is nervous or secretive (or you're busy), you might miss the scratching and just notice short, thinned-out patches of fur, and sometimes scabs or raw wounds. Nibbling loose fur also increases the risk of vomiting hairballs. Either way, it's time to contact your vet.

Fight those fleas

There's no quick fix, so before reaching for the flea treatment, get advice from your vet. If you use the wrong product or fail to take other simple measures, you'll be fighting a losing battle. Simply waiting for something to change risks missing other potential causes of the scratching, while those claws could be doing some serious damage.

What should I do?

In the moment:

- **Notice your cat's** body language, especially their ears (see pages 12–13). Do they seem distressed, agitated, or uncomfortable?
- **Could it be fleas?** Indoor cats can still get fleas—from people, other pets, rodents, or items entering the home—and central heating helps them thrive all year. As well as being irritating, fleas can cause anemia and transmit diseases. Do the flea test: sprinkle debris from your cat's favorite sleep spots onto a piece of toilet paper and add a few drops of water. If it turns red-brown, it's digested cat blood, a flea's meal of choice—see box, below.

In the longer term:

- **Get a vet check**, especially if either the skin or fur looks abnormal or the behavior persists—other causes could be allergies, parasites, ear and skin infections, illness, or a side-effect of medication.
- **Trim their claws** if they don't go outside. This reduces damage in the short-term while other treatments kick in. Book a "pawdicure" at the vet, or do it yourself if you know how.
- **Keep up-to-date** with the right flea treatment for your cat (see left). Never use dog flea products on a cat, as they can be lethal.

My cat hates being picked up

My last cat loved being cradled like a furry baby, but my current cat hates it. They're such a softy the rest of the time, but they go as stiff as a board, and then rip me to shreds trying to get away.

What's my cat thinking?

Your cat's frantic response suggests fear and frustration because that well-meaning hug is definitely taking them outside of their comfort zone. If they could talk, they'd probably say, "Put me down this instant!"

From your cat's perspective, out of the blue, you've exerted physical control over them, which prevents them from fleeing any situation that may arise that they find unsettling, scary, or painful. It's the kitty equivalent of being shackled. If your cat wasn't picked up as a kitten (see pages 18–19), their response now might be due to a fear of the unknown, or maybe it's given them a flashback of a bad experience earlier in life—or of their last trip to the vet.

Loss of control

Cats are small and vulnerable to predators from above, such as coyotes or birds of prey, so being lifted off the ground can be seen as a threat in itself. Your cat's only objective is to retain control and make a hasty exit—which requires four feet firmly on the ground.

What should I do?

In the moment:

- **If it upsets your cat**, don't do it! Let them go and leave them alone.

In the longer term:

- **If this is new behavior**, rule out illness or pain—feeling under the weather, or having a recent injury, or a nagging backache or toothache can make anyone tetchy. Cue vet check.
- **Assess the quality** of your cat's world (see pages 46–47, 138–139, and 182–183). Could they be on edge due to some underlying frustration or anxiety?
- **Decide whether** you're going to respect that cuddles give your cat the heebie-jeebies and move on, or whether you're committed to training them to allow—and even enjoy—being picked up. Learning to be at ease with handling makes vet trips, nail trims, travel, and giving medication less stressful.
- **Think of all the nonfood** ways to show your love on your cat's terms. Focus on those instead of meeting your own needs with enforced hugs.

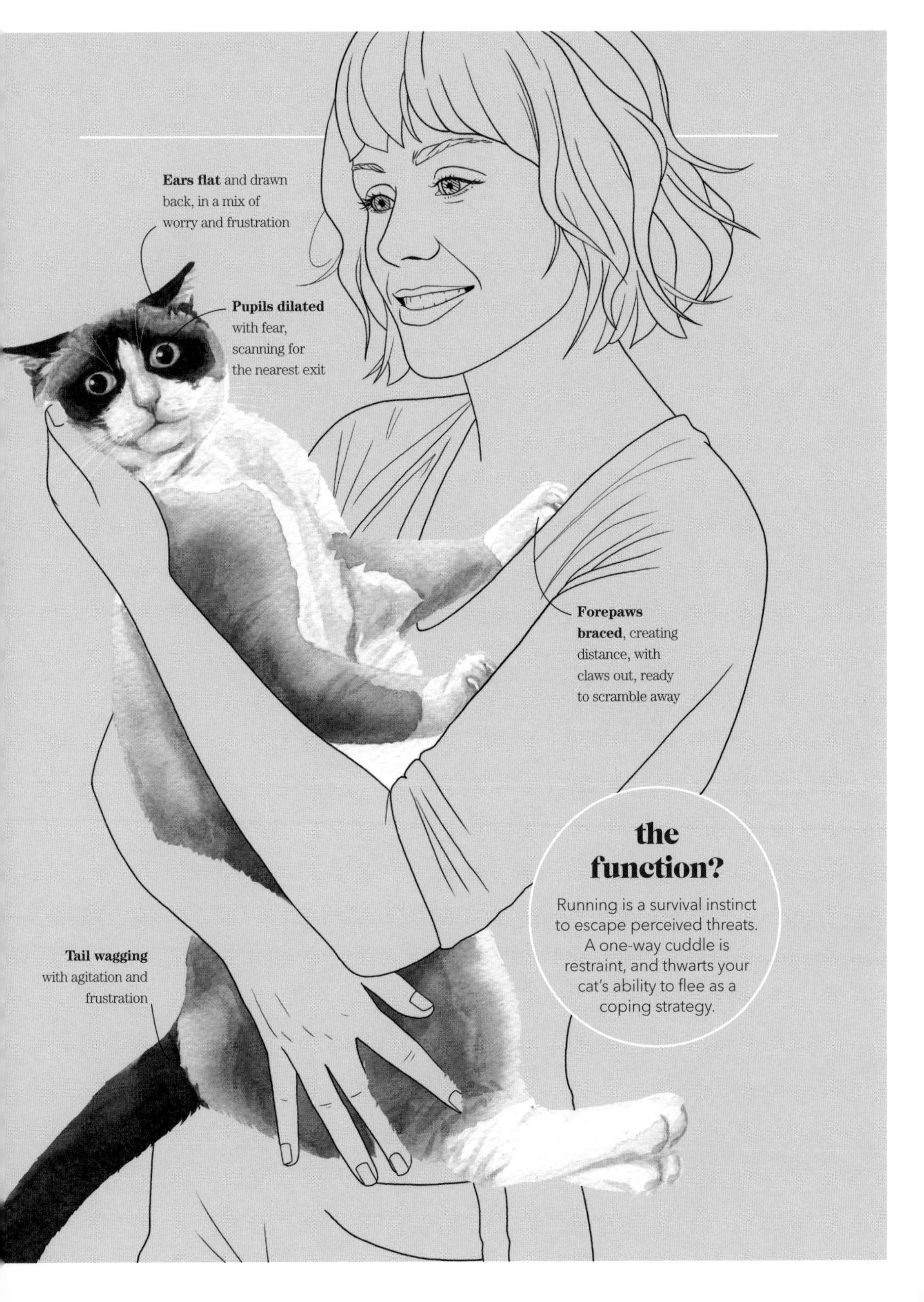

the function?

Running is a survival instinct to escape perceived threats. A one-way cuddle is restraint, and thwarts your cat's ability to flee as a coping strategy.

ADVANCED CAT WATCHING

Giving cats a bad name

The labels commonly given to cats are loaded with connotations that influence our thinking and stop us from seeing the root cause of their behavior. Cats are not furry human babies or small dogs; beneath their designer fur coats, they're very much in tune with their inner wildcat—so the only label that should stick is "cat"!

A nasty bully

Cats are territorial, not nasty. Their wildcat ancestors lived solitary lives, and pet cats are still hardwired to communicate from a distance and to see other cats as threats or rivals. It's completely unnatural for them to be living in close quarters with other cats they don't know. But some are just far more easy-going about being put in this situation and set the bar too high for the rest.

Acting out of spite

Cats feel reassured by familiar smells and sights and get anxious when things change, especially within their core territory. They're not defying you because you've pulled their favorite rug out from under them and updated your sofa, but it's made them feel insecure and confused, with strange smells in their usual snoozing spot. With the comforting scent of home wiped out, drastic measures were called for (see pages 14–15).

Naughty at the vet

Lashing out at the vet doesn't make a cat naughty, bad, or evil. They're scared, confused, and potentially in pain. If you were woken abruptly from your sleep, kidnapped, caged, and stuck in a room full of predators, only to be dragged onto a slippery platform by a needle-wielding stranger, you'd protest in any way you could, too!

Pet cats are cruel

The truth is, they're confused, not cruel. All cats are predatory–as carnivores, they have to hunt live prey to survive. They're also opportunistic feeders, so the desire to kill isn't linked to hunger. A pet cat has a wildcat's instinct to kill, but might not know what to do with the prey afterward. It's more to do with a plentiful supply of meals and an under-stimulating home environment than a thirst for carnage.

Cats are so aloof

Physically, cats can't express themselves facially to the same extent that humans or dogs can. This doesn't make them suspicious or calculating, and has nothing to do with a lack of feelings or personality. It's simply about evolution and anatomy–facial expressions aren't much use when you've evolved to communicate from a distance. Cats have other ways to convey their emotional status (see pages 12–17), but for some humans it's just easier to mislabel a cat as unfriendly than to try to understand them.

My old cat cries at night

It sounds alarming so I leap out of bed, but they're just sitting with a vacant expression. Are they losing their marbles?

What's my cat thinking?

Let's be honest, old age isn't kind to either the body or the brain. Functions such as the sleep–wake cycle and memory deteriorate, meaning that learned behaviors and routines are forgotten. The five senses that cats rely on to gather information about their surroundings start to wane, too. This can be a big deal, especially at night when it's dark, which can intensify confusion and disorientation. Add to the mix diseases that cause thirstiness, midnight munchies, and more toilet stops, and it's easy to see why your cat could be feeling insecure and anxious. They're sending out an SOS, so don't ignore it.

Glazed eyes and dilated pupils indicate an eye check is necessary

Wailing meows are a cry for help

the function?

Meowing in this context is a cry for help. Your cat is feeling bewildered or vulnerable and could need medical care. Either way, they need your assistance.

What should I do?

In the moment:

- **Make sure your cat** is not in immediate danger or acute distress.
- **Check they have** access to water, litter box, and a warm place to rest.
- **Stay calm** and don't do anything that might reinforce this behavior (see pages 108–109).

In the longer term:

Think about all the ways in which you could support your cat's aging body. Try the following:

- **Leave a lamp on** or use plug-in nightlights, so darkness doesn't add to your cat's feelings of confusion.
- **Increase the number** of litter boxes; food and water stations; and warm, cozy spots, in case your cat has become forgetful. Ensure they are easy to access and add heat pads to beds to ease old, arthritic limbs.
- **Avoid changing home** layouts or routines, as it can throw your cat off-kilter. It's harder to teach an old cat new tricks, so keep things as familiar as possible and don't clean away all those recognizable smells.

See your vet!

If late-night distress becomes a regular occurrence, your cat needs to be checked by the vet. Advances in feline medicine mean cats can live happily, well into their twenties, with the right care—but your vet can't catch disease early if you take a wait-and-see approach, so make that call.

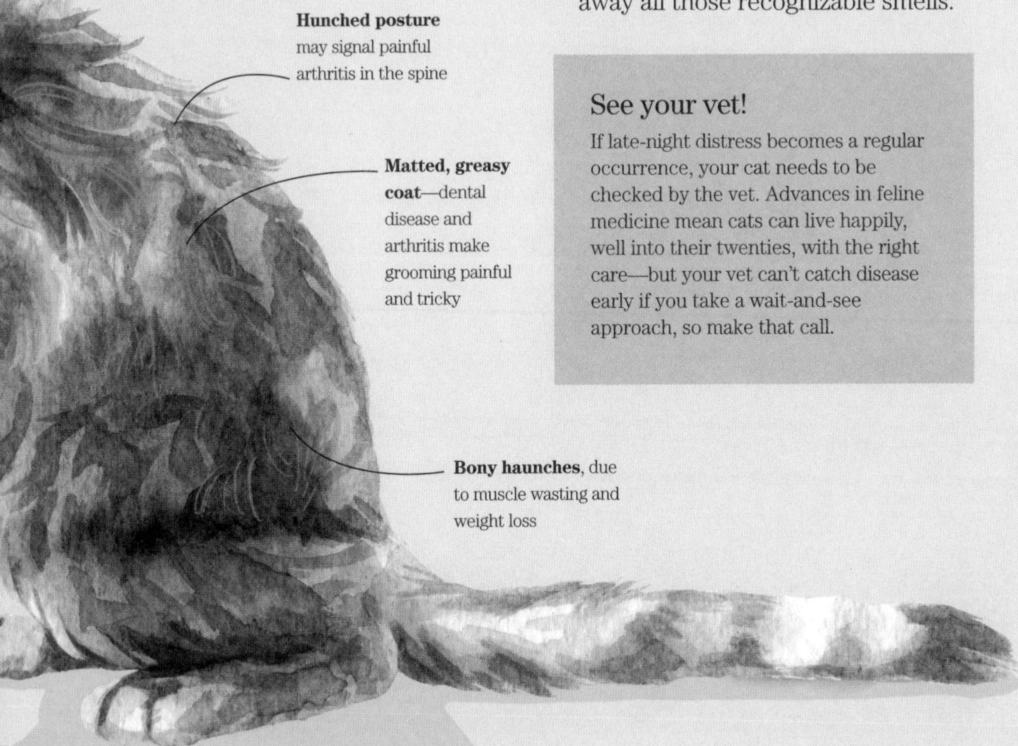

SURVIVAL GUIDE

Positive play

Play is an outlet for a cat's natural predatory urges. Pretend hunting, with toys that give them a thrill and no gory aftermath, keeps their mind and body fit and reduces stress and boredom.

1

Keep "prey" real

The perfect toy "prey" looks, feels, moves, and sounds like the real deal. Seek out mouse-size furry or feathered toys that squeak or tweet. If it falls apart during a supervised attack, all the better.

2

Keep it interesting

Patting balls and bunny kicking rat-size toys is fun, but the best games involve you, such as fetch and fishing-rod toys. Vary what you play with, and marinate toys in a jar of catnip to increase their appeal.

3

Keep it safe

Cats literally go wild while playing, so stay away from teeth and claws using rod toys. Store toys away safely when unsupervised and check them before use to make sure they are intact and safe.

4

Keep it positive

Play helps keep cats fit and happy, releasing pent-up energy and stress, improving confidence, and reducing undesirable behaviors and the risk of obesity. Chasing the uncatchable–lasers, bubbles, or simulated prey on apps or TV–can be frustrating. End some games with a toy or treat to give them the satisfaction of a capture.

5

Keep play real

Replicate hunting rituals, ideally at dusk and dawn, with several bursts of play in a half-hour session, followed by a meal. Mimic the prey sequence (see pages 70–71), alternating flying, landing, wriggling, and pausing as you move the toy away from the hunter. Offer tunnels, paper bags, and boxes for stalking, rustling, and hiding.

My cat tries to hump me

I've known dogs to be problem humpers, but my neutered cat does it, too—to furniture, other pets, and, worst of all, parts of my body! Why is my kitty so into heavy petting?

What's my cat thinking?

Both male and female cats, even when they have been neutered, still have the capacity to become aroused, but often it's not the potential for sex that gets them going. The action of humping triggers the brain to release a cocktail of "cuddle" and "bliss" hormones—oxytocin, serotonin, and dopamine—and these endorphins bring a sense of pleasure and control.

Humping is more common in stressed cats who aren't free to follow their wildcat rhythm and instincts, or are lacking the right habitat or regular human affection. It's also seen in cats suffering from urinary tract disease, so rule this out with a vet check.

Fantastic phallic facts

It's completely normal for male cats to become aroused when grooming the genital area. An unneutered male's penis is lined with barbs, which lock into the female during mating (ouch!) and stimulate egg release. These barbs disappear after neutering. And did you know, unlike humans, some cats have a bone inside their penis? Fascinating!

What should I do?

In the moment:

- **Don't overreact**—shouting or jumping around will alarm your cat and only reinforce this stress-busting behavior.
- **Don't try to move** them off while they're in the throes of passion—you'll be safer from flailing claws if you ride it out (so to speak).
- **Look for clues**—could the motivation for humping be frustration or anxiety? What set them off? Was it a new scent, or being deprived of something they want or need? Was there conflict with another pet?

In the longer term:

- **Redirect their energy**—whenever you spot the signs of imminent humping, divert them toward something more acceptable than a part of your body—such as a cushion or a soft toy.
- **Swap the sexual endorphins** rush for a predatory one by engaging them in exciting play with a wand toy (see pages 182–183).

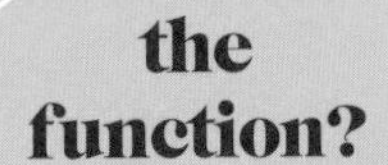

the function?

Apart from the obvious sexual function, humping also soothes anxiety and tension, as it releases hormones that boost mood and relieve stress.

Hips straddling the object of affection and pumping back and forth

Ears slightly back and flat, indicating frustration

Jaws clamped—holding the "partner" still

Back feet treading up and down, a part of the mating ritual

Index

Breed index

Resources

Further reading

Cat Sense John Bradshaw

Feline Stress and Health (ISFM Guide) edited by Sarah Ellis and Andy Sparkes

The Trainable Cat John Bradshaw and Sarah Ellis

Online

www.thecatvet.co.uk
The Cat Vet's own expert online resources will give you the skills and tools you'll need to #ThinkLikeACat and keep their body and mind in optimal condition from the comfort of your home.

www.icatcare.org
International Cat Care is a charity headed by feline vets committed to feline health and welfare.

www.aspca.org/pet-care/animal-poison-control
The Animal Poison Control Center (APCC) division of the American Association of the Prevention of Cruelty to Animals offers advice on common toxins.

https://cfa.org/
The Cat Fanciers' Association offers information and advice to help people choose the right cat for their circumstances and lifestyle.

www.tica.org
The International Cat Association (TICA)—the world's largest genetic registry of pedigreed pet cats and kittens.

Help with your cat's behavior

If you're wondering whether to seek further help with your cat's antics, avoid "Dr. Google" and contact a certified Cat Friendly Vet (see below) to rule out a medical problem. Subtle changes in behavior can be an early sign of illness, so diagnosing and treating any issues promptly will help keep your cat happy and healthy. You might just need to make some subtle #ThinkLikeACat tweaks to your home, routine, or mindset, which your vet can also help with.

Don't waste time trying to guess what the problem is—that's your vet's job. Whether you've spotted a new behavior or noticed that an old one has changed or ceased, your only task is to book a checkup with the vet. They're the best people to give your cat any medical help they need, and give you the support and resources you need—whether that's suggestions for improving your cat's habitat, help getting pills down them, or recommendations for a trusted cat sitter. Sometimes cats need to be referred to a certified feline behaviorist or counselor. The world of kitty therapy is poorly regulated, but your vet will be able to recommend someone with cat expertise whom they trust. This might be a specialist behavior vet (akin to human psychiatrist) for serious or complex cases, or a feline psychologist for more straightforward issues.

Finding a certified Cat Friendly Vet

www.catfriendly.com/find-a-veterinarian
US and Canada: AAFP (American Association of Feline Practitioners)

www.catfriendlyclinic.org
Rest of the world: ISFM (International Society of Feline Medicine)

You'll increase your chances of finding a local certified Cat Friendly Vet who does home visits by entering certain keywords into your Internet search engine: "cat vet" and "your local area," with phrases such as "home visiting," "mobile," and "house call."

Finding a cat sitter

Ask your vet for a recommendation—some practice staff pet sit. Interview candidates and look out for signs that they understand and like cats (see pages 90–91).

www.narpsuk.co.uk
UK: National Association of Pet Sitters and Dog Walkers

www.petsit.com/locate
US & Canada: Find a Certified Professional Pet Sitter (CPPS) registered with Pet Sitters International (PSI)

References

14 Communicating with scent
Vitale Shreve K R, Udell M A R. Stress, security, and scent: The influence of chemical signals on the social lives of domestic cats and implications for applied settings. *Appl Anim Behav Sci* 2017; 187: 69–76 https://doi.org/10.1016/j.applanim.2016.11.011

16 Communicating with sound
McComb K, Taylor A M, Wilson C, Charlton B D. The cry embedded within the purr. *Curr Biol* 2009; 19 (13): 507–508. https://doi.org/10.1016/j.cub.2009.05.033

40 My cat goes crazy for catnip
Bol S, Caspers J, Buckingham L, *et al.* Responsiveness of cats (*Felidae*) to silver vine (*Actinidia polygama*), Tatarian honeysuckle (*Lonicera tatarica*), valerian (*Valeriana officinalis*) and catnip (*Nepeta cataria*). *BMC Vet Res* 2017; 13: 70. https://doi.org/10.1186/s12917-017-0987-6

42 My cat thinks they're a cow
Franck A R, Farid A. Many species of the Carnivora consume grass and other fibrous plant tissues. *Belg J Zool* 2020; 150: 1–70. https://doi.org/10.26496/bjz.2020.73

68 My cat chatters at the birds
de Oliveira Calleia F, Rohe F, Gordo M. Hunting strategy of the margay (*Leopardus wiedii*) to attract the wild pied tamarin (*Saguinus bicolor*). *Neotropical Primates 2009*; 16 (1): 32–34. https://doi.org/10.1896/044.016.0107

74 My cat kneads and drools on me
Matulka R A, Thompson L, Corley. Multi-Level Safety Studies of Anti Fel d 1 IgY Ingredient in Cat Food. *Front Vet Sci* 2020; 6: 477 https://doi.org/10.3389/fvets.2019.00477

116 My cat sucks and chews odd things
Kinsman R, Casey R, Murray J. Owner-reported pica in domestic cats enrolled onto a birth cohort study. *Animals (Basel)* 2021; 11 (4): 1101. https://doi.org/10.3390/ani11041101

136 My cat is a counter surfer
Wells D L, McDowell L J. Laterality as a tool for assessing breed differences in emotional reactivity in the domestic cat, *Felis silvestris catus. Animals (Basel)* 2019; 9 (9): 647. https://doi.org/10.3390/ANI9090647

148 My cat is so clingy and needy
Mira F, Costa A, Mendes E, *et al.* A pilot study exploring the effects of musical genres on the depth of general anesthesia assessed by hemodynamic responses. *J Feline Med Surg* 2016; 18 (8): 673–678. https://doi.org/10.1177%2F1098612X15588968

150 My cat hates going to the vet
Hampton A, Ford A, Cox R E, *et al.* Effects of music on behavior and physiological stress response of domestic cats in a veterinary clinic. *J Feline Med Surg* 2020; 22 (2): 122–128. https://doi.org/10.1177/1098612X19828131

184 My cat tries to hump me
Tobón R M, Altuzarra R, Espada Y, *et al.* CT characterisation of the feline os penis. *J Feline Med Surg* 2020 Aug; 22 (8): 673–677. https://doi.org/10.1177/1098612X19873195

Acknowledgments

Author's acknowledgments

This book was a real labor of love, made possible by some amazing people to whom I'd like to give thanks:

Andy, for believing in me and offering endless support and cups of tea. It's not easy living with a purr-fectionist who's juggling two babies, three cats, and deadlines! To my family and in-laws, for their love and encouragement—especially my Mum and Dad, for teaching me kindness, the value of hard work by example, and never to give up on my dreams … if it's not okay, then it's not the end!

To my fellow cat vet Vanessa, my dearest friend, thanks for sharing the journey down the road less traveled to vet school. Against all odds, we did it! To Dr. Sue Beetson and all the other vets, nurses, and path lab geeks who believed in a determined young Jo. You'll never know quite what your support meant to me, when you gave me the chance as a teenager to spend my Saturdays up to my armpits in cat pee and poo!

Thank you to all my wonderful clients who've invited me into their cats' homes, shared their cats' antics and stories, and, most importantly, entrusted me with the care of their beloved furry family members. I'm eternally grateful.

To the DK team, especially Rona, Zia, Karen, Dawn, Louise, and Marianne—such lovely fellow cat people, which made it so easy to work with you. Thanks for allowing me the privilege of unraveling the feline mind in print. To Mark, thank you for bringing my scribbled stick-figure cats to life with such beauty and cattitude.

Last but not least, to my beautiful cats, past and present, to whom I owe the biggest debt of all. You have given me what no vet degree could—firsthand experience of being a cat lover. Because you can only truly understand the challenges of a multi-cat household, the struggle of medicating and caring for a sick cat, or the heartache of saying your last goodbyes when you've been there yourself. The highs and lows of your little lives have helped me to share what I've learned with so many people and their cats. I hope that, if you could read, you'd approve of what I've written.

Publisher's acknowledgments

DK would like to thank Marie Lorimer for indexing and John Friend for proofreading.

Disclaimer

About the author

Dr. Jo Lewis is an award-winning British veterinarian with more than 25 years' experience studying and working with cats. She is an American Association of Feline Practitioners (AAFP) Certified Cat Friendly Veterinarian and a member of the International Society of Feline Medicine (ISFM).

After graduating from vet school with First-Class Honors, Jo quickly realized how stressed cats and their owners became when visiting the vet, so she made it her business to do something about it. She founded The Cat Vet and the UK's first home-visiting vet clinic dedicated entirely to cats, offering them compassionate, stress-free vet care and expert cat advice in the comfortable surroundings of their own home. Her next #ThinkLikeACat project is developing online courses and workshops to teach people how to keep their cats happy and healthy in both mind and body.

During her career, Jo has worked alongside other cat vets at the world-class Oxford Cat Clinic, and airside as a Government Vet Inspector at London's busy Heathrow Airport. A self-confessed lab geek, she spent years as a clinical pathology consultant advising vets in the UK about anything from tabbies to tigers.

Jo cares passionately about the welfare of all creatures, great and small. While living in Australia, she volunteered to help with dolphin and snake conservation projects, and she has spent many a sleepless night hand-raising orphaned kittens. Never off-duty, she even rolled up her sleeves during her honeymoon to help a local charity trap a badly injured, elusive street cat. Currently, Jo is purr-sonal assistant to a rescue Siamese, two moggies, and two little cat people.

See more about Jo and her work at **thecatvet.co.uk**.

About the illustrator

Mark Scheibmayr is an illustrator specializing in pet portraiture, based in Toronto, Canada. A proud co-parent to a rescue dog, he has also owned and been a friend to many cats. In addition to illustrating several books for DK, Mark has done editorial work for Chapters Indigo, DK Travel, and DobbernationLOVES. Other illustration work includes campaigns for the Town of Milton and exhibition design for Markham Museum. See more of his work at **markscheibmayr.com**.

Image credits

The publisher would like to thank the following for their kind permission to reproduce their photographs: (Key: a-above; b-below/bottom; c-center; f-far; l-left; r-right; t-top). Cover image: **Depositphotos Inc:** Photocreo 19 **Dreamstime.com:** Bogdan Sonyachny. 20 **Alamy Stock Photo:** Linda Kennedy. 46 **Shutterstock.com:** Anurak Pongpatimet. 62 **Dreamstime.com:** Nils Jacobi. 90 **Getty Images / iStock:** ablokhin. 96 **Dreamstime.com:** Famveldman. 112 **Getty Images / iStock:** chendongshan. 126 **Getty Images / iStock:** Valeriya. 138 **Dreamstime.com:** Insonnia. 152 **Dreamstime.com:** David Herraez. 156 **Dreamstime.com:** Fotosmile. 170 **Dreamstime.com:** Daria Kulkova. 182 **Shutterstock.com:** Dora Zett. 192 **Dr Jo Lewis** (tr) **Mark Scheibmayr** (clb)